Michael Lublow

Atomic Terraces, Nanostructures and Fractals on Silicon

Michael Lublow

Atomic Terraces, Nanostructures and Fractals on Silicon

Chemistry and Photoelectrochemistry at the Silicon/Silicon Oxide/Electrolyte Phase Boundaries - A Surface Analytical Characterization

Südwestdeutscher Verlag für Hochschulschriften

Impressum/Imprint (nur für Deutschland/ only for Germany)
Bibliografische Information der Deutschen Nationalbibliothek: Die Deutsche Nationalbibliothek verzeichnet diese Publikation in der Deutschen Nationalbibliografie; detaillierte bibliografische Daten sind im Internet über http://dnb.d-nb.de abrufbar.

Alle in diesem Buch genannten Marken und Produktnamen unterliegen warenzeichen-, marken- oder patentrechtlichem Schutz bzw. sind Warenzeichen oder eingetragene Warenzeichen der jeweiligen Inhaber. Die Wiedergabe von Marken, Produktnamen, Gebrauchsnamen, Handelsnamen, Warenbezeichnungen u.s.w. in diesem Werk berechtigt auch ohne besondere Kennzeichnung nicht zu der Annahme, dass solche Namen im Sinne der Warenzeichen- und Markenschutzgesetzgebung als frei zu betrachten wären und daher von jedermann benutzt werden dürften.

Verlag: Südwestdeutscher Verlag für Hochschulschriften Aktiengesellschaft & Co. KG
Dudweiler Landstr. 99, 66123 Saarbrücken, Deutschland
Telefon +49 681 37 20 271-1, Telefax +49 681 37 20 271-0
Email: info@svh-verlag.de
Zugl.: Cottbus, BTU, Diss., 2009

Herstellung in Deutschland:
Schaltungsdienst Lange o.H.G., Berlin
Books on Demand GmbH, Norderstedt
Reha GmbH, Saarbrücken
Amazon Distribution GmbH, Leipzig
ISBN: 978-3-8381-1776-8

Imprint (only for USA, GB)
Bibliographic information published by the Deutsche Nationalbibliothek: The Deutsche Nationalbibliothek lists this publication in the Deutsche Nationalbibliografie; detailed bibliographic data are available in the Internet at http://dnb.d-nb.de.

Any brand names and product names mentioned in this book are subject to trademark, brand or patent protection and are trademarks or registered trademarks of their respective holders. The use of brand names, product names, common names, trade names, product descriptions etc. even without a particular marking in this works is in no way to be construed to mean that such names may be regarded as unrestricted in respect of trademark and brand protection legislation and could thus be used by anyone.

Publisher: Südwestdeutscher Verlag für Hochschulschriften Aktiengesellschaft & Co. KG
Dudweiler Landstr. 99, 66123 Saarbrücken, Germany
Phone +49 681 37 20 271-1, Fax +49 681 37 20 271-0
Email: info@svh-verlag.de

Printed in the U.S.A.
Printed in the U.K. by (see last page)
ISBN: 978-3-8381-1776-8

Copyright © 2010 by the author and Südwestdeutscher Verlag für Hochschulschriften Aktiengesellschaft & Co. KG and licensors
All rights reserved. Saarbrücken 2010

Contents

Introduction 3

1. Fundamental aspects 7

1.1 Chemical, structural and electronic properties of the SiO_2/Si interface _____ 7

1.1.1 Silicon and silicon dioxide bulk properties... 7
1.1.2 Properties of the SiO_2-Si interface 9
1.1.3 Stress and strain at the interface 11

1.2 Competing electronic and (photo-)electrochemical processes at the reactive semiconductor-electrolyte interface_____ 13

1.2.1 The Marcus theory of single electron transfer 13
1.2.2 Energy levels in semiconductors and redox systems 17
1.2.3 Semiconductor (photo-)corrosion 21

1.3 Self-organization phenomena at the silicon/electrolyte interface _____ 24

1.3.1 Dynamical systems 24
1.3.2 Self-organization phenomena at silicon electrodes 25

2. Experimental methods and procedures.. 28

2.1 Brewster-angle analysis _____ 28

2.1.1 The dielectric function 28
2.1.2 Brewster-angle analysis of multi-layer systems 33

2.2 *In situ* Brewster-angle reflectometry of (electro-)chemical conditioned silicon surfaces_____ 40

2.2.1 Experimental arrangement.. 40
2.2.2 The linear approximation of the reflectance 41

2.3 Photoelectron spectroscopy using synchrotron radiation _____ 42

2.3.2 Principles of photoelectron excitation 42
2.3.2 The application of synchrotron radiation 48

2.4 Photoelectron emission microscopy _____ 52

2.4.1 Experimental arrangement.. 52
2.4.2 Contrast in photoelectron emission microscopy 53

2.5 Atomic force microscopy _____ 54

2.6 Scanning electron microscopy _____ 57

2.6.1 Experimental arrangement.. 57
2.6.2 Depth of field, chemical and spatial resolution 61

3. Results and discussion.. 65

3.1 Identification of a sub-surface stressed silicon layer 65

3.1.1 Introductory remarks 65
3.1.2 *Ex situ* Brewster-angle analysis: in loco etching results 66
3.1.3 *In situ* Brewster-angle reflectometry: real-time monitoring results 78
3.1.4 Morphological and chemical optimization of Si(111)-1x1:H 91
3.1.5 Synopsis: chemical and structural properties of the stressed interfacial region ... 99

3.2 Horizontal nanostructure formation by photoelectrochemical conditioning 100

3.2.1 Alignment effects and shaping of nanostructures in the divalent dissolution region 101
3.2.2 Model considerations for the self-organized and engineered nanostructure formation 115
3.2.3 Structural changes at the Si(111) interface during anodic oscillations 121
3.2.4 Summary: *in situ* controlled self-organized nanostructure formation 131

3.3 Self-organized propagation of fractal silicon microstructures in concentrated NH_4F 133

3.3.1 Background on fractals and macropores 133
3.3.2 Experimental results for Si(111), (110), (100) and (113) surfaces 134
3.3.3 Influence of the photon flux on the surface topographic degree of order 150
3.3.4 On the role of interfacial stress on fractal structure propagation 154
3.3.5 Numerical simulation of the microstructure formation 157
3.3.6 Summary and comparison of lateral with vertical macropore formation 162

Summary 166

Appendices 168

A.1 Multi-layer analysis of Brewster-angle data 168

A.2 Numerical procedure for simulation of the stress-induced propagation of fractal micro- and nanostructures 172

Acknowledgments 176
References 177

Introduction

The silicon dioxide / silicon interface (SiO_2/Si) is one of the most intensively investigated interfaces in semiconductor technology since Frosch and Derick reported the outstanding passivation properties of the SiO_2 layer in 1957 [1]. With invention of the metal-oxide-semiconductor field-effect transistor (MOSFET) in 1960 [2], the application of SiO_2 as gate oxide appeared most promising due to the low density of electronic interface states and determined therefore the manufacturing process of integrated circuits for all the decades that would follow. The importance of the oxide layer and its interface towards bulk silicon is therefore comparable to silicon itself and modern technical culture, termed as "silicon age", relies equally on both. With the beginning Millennium, the detailed inspection of the structural, chemical and electronic properties of this interface has not ceased. 'The oxide gate dielectric: do we know all we should?' entitles suggestively a review article from 2005 [3], emphasizing that research still has to focus on the gate oxide to meet the requirements for future device fabrication. It is true that new materials have to be employed in order to fulfill the objectives of the semiconductor industry roadmap. Intel's 45 nm technology consequently integrates high-κ dielectrics into advanced MOS structures [4]. But refined experimental techniques as well as advanced theoretical approaches provide continuously deeper insight into the SiO_2/Si interface and accumulate knowledge that will possibly apply for future gate materials and their interfaces, too.

While SiO_2 in nanoelectronics is being gradually substituted for new oxides, its role as global or local, micro- or nano-stressor of the underlying silicon substrate, as investigated in the work here, points possibly to new engineering strategies in the rapidly growing field of Micro- and Nanoelectromechanical Systems (MEMS / NEMS) [5]. As an unprecedented paradigm in both science and technology, micro- and nanostructures at surfaces of solids are presently subject of a combined and interdisciplinary approach in physics, chemistry, biology and (ultra-)precision engineering. Possible applications are ranging from sensing devices in chemistry, pharmaceutics and biology, photovoltaic and photocatalytic applications to micro-mirrors in optical variable devices (OVD) and even fully equipped micro-labs for interplanetary research. Moreover, in multi-scale system design, solutions for the integration of variable length-scale components on single devices are sought that cover the full range from the millimeter to the nanometer scale.

In this work, micro- and nanostructures at the SiO_2/Si interface are investigated. As one of the improved experimental techniques, alluded to above, *ex situ* Brewster-angle analysis and *in situ* Brewster-angle reflectometry are extensively employed in combination with microscopic and photoelectron emission techniques. Chemical, optical and topographical properties are investigated for structures buried beneath top surface oxide layers as well as for time-dependent micro- and nanostructure formation evolving during photoelectrochemical conditioning in ammonium fluoride containing solutions (NH_4F). The variety of observed topographies ranges from randomly to symmetrically aligned surface patterns with pronounced dependence on the underlying surface lattice properties. Upon self-organized electrochemical dissolution, oxide induced vertical and lateral stress gradients at the interface will be identified as one of the most determinant feed-back mechanisms for shape and propagation of the micro- and nanostructures.

Stress fields at the interface are resulting from both the lattice mismatch of Si and SiO_2 as well as incorporation of oxygen atoms into the silicon bulk [6]. Generally, interface stress is difficult to assess directly for it is often superimposed by dielectric, topographical and mechanical (bulk) properties of the two adjacent materials, the silicon substrate and the oxide layer. Separation of the respective contributions in VIS/UV optical [7] or high resolution X-ray diffraction experiments [8] is therefore challenging: data interpretation often depends strictly on the validity of initial model considerations. For instance, in a first *ex situ* application of the highly surface sensitive method of Brewster-angle analysis, stress at the interface of native oxide covered Si(111) is only deducible by *post-experimental* multi-layer analysis of the optical data, measured for the successive etch-back of the ultra-thin oxide layer. Conversely, *in situ* monitoring by Brewster-angle reflectometry allows an almost unadulterated observation of the accelerated dissolution of the stressed interface during wet chemical etching. It will be comprehensible from these experiments that a transition layer with stressed/strained atomic bonds increases the nominal thickness of the SiO_2/Si interface by a few nanometers. This knowledge is subsequently applied to the analysis of silicon micro- and nanostructures prepared by photoelectrochemical conditioning in varying NH_4F solutions.

In diluted NH_4F, the evolution and shaping of regular nanostructures will be controlled by application of *in situ* Brewster-angle reflectometry. A new light intensity variation technique is developed which allows subsequent shaping of prefabricated nanostructures. The resulting improved aspect ratio of the structures, as an outcome of selective oxidation, is qualitatively interpreted in terms of reinforced oxide growth along stressed atomic silicon bonds. These structures are then compared to the morphologies at the silicon interface which evolve during repeated photocurrent oscillation cycles at increased anodic potentials.

In concentrated NH$_4$F, fractally branching etch domains, extending from the sub-micrometer to the millimeter range and deeply etched into the substrate, are observed. High-Resolution Scanning Electron Microscopy (HR-SEM) clearly proves the dependence of the domains on the respective surface lattice symmetries. The domains encircle furthermore ensembles of smoothly polished, slow etching microfacets which makes model development complicated. It will be one of the important conclusions, corroborated by the experimental findings, that outer contours and inner topographies evolve, to some extent, independently from each other. Quantitative in-plane stress analysis confirms that lateral stress gradients promote the regular propagation of the microcracks and accounts for many of the novel effects such as transition from well aligned to aperiodic (random) structures, scaling effects upon light intensity variation as well as mutual repelling of approaching individual structures.

In the introductory **Chapter 1, Fundamental Aspects**, structural, chemical and electronic properties of the SiO$_2$/Si interface are summarized. The single electron transfer process across the semiconductor/electrolyte interface is described according to the Marcus-Gerischer approach. Chemical and electrochemical routes of silicon dissolution in NH$_4$F are shown. Finally, a brief introduction into self-organized dynamic systems is presented with particular attention to those phenomena observed at the silicon/electrolyte interface.

Chapter 2, Experimental Methods, outlines fundamental aspects of the applied methods as important for the understanding of the experimental results. Details of the optical methods, *ex situ* Brewster angle analysis and *in situ* Brewster-angle reflectometry, are presented with application to optical multi-layer analysis. Chemical analysis by Photoelectron Spectroscopy is described with particular emphasis on soft X-ray application using synchrotron radiation at Bessy II, Berlin-Adlershof. Scanning electron microscopy and Atomic Force Microscopy (AFM) are finally introduced as microscopic techniques. It should be noted that proper chemical silicon surface preparation in this work is not a prerequisite but a major result despite the fact of innumerable publications in this area. It was motivation to provide an improved method of silicon surface preparation that allows simultaneously large sampling numbers and optimized surface quality.

In **Chapter 3, Results and Discussion**, the selective *in loco* etching of ultra-thin oxide layers, analyzed by BAA and AFM, is presented. Oxide thickness analyses are performed in comparison to results obtained by surface sensitive photoelectron spectroscopy. The unexpected roughening of Si(111) surfaces during etching in concentrated ammonium fluoride solution is subsequently investigated by *in situ* BAR, resulting in a model description of the etching process from oxide removal to silicon bulk etching.

In the non-oscillatory regime of photoelectrochemical silicon dissolution, a procedure is introduced for silicon nanostructure formation and shaping, controlled by *in situ* BAR. The specific behavior of n-type Si photoelectrodes, immersed in diluted NH_4F solutions, permits manipulation of the dissolution reaction by light intensity variation and results in increased aspect ratios of the structures. In the oscillatory regime, varying oxide thicknesses and interface topographies are assessed by *ex situ* BAA and AFM.

Fractal microstructures, produced by photoelectrochemical dissolution in concentrated NH_4F, are finally presented. Based on chemical and topographical surface analyses, a stress induction model is proposed to account for formation and propagation of the extended crack patterns. The relation between structure symmetries and crystal lattice properties are discussed on the basis of analytical and numerical computations.

The **Appendices** finally are detailing the mathematical and computational procedures employed for optical multi-layer analysis as well as for simulation of propagating microcracks.

1. Fundamental aspects

1.1 Chemical, structural and electronic properties of the SiO$_2$/Si interface

1.1.1 Silicon and silicon dioxide bulk properties

Silicon atoms are arranged at the corners of a tetrahedron with a center silicon atom [9]. The unit cell length which defines the lattice constant of the material is 5.43 Å. The covalent bond length to the nearest neighbor is 2.38 Å. Due to long-range order, these unit cells form the whole crystal by repeated cell alignment and the crystal exhibits a face cubic centered (fcc) symmetry [10]. In Fig. 1-1, the unit cell as well as selected viewing directions towards the (100) and (111) plane, as the most important surface orientations, are shown. The electronic configuration of silicon is composed of filled K and L shells (L1-L3) as in the element Neon plus four valence electrons in the 3s^2 and 3p^2 state which transform to so-called sp^3 hybrid orbitals for energetic reasons, i.e., the energy of the p-state electrons is lowered while the corresponding energy of the s-state electrons is raised, resulting thus in an energy gain for the sp^3 system. In the extended crystal, the overlap of the atomic wave functions leads to energy splitting and eight energy bands are forming: four valence bands and four conduction bands which are, at T = 0 K, completely filled and, respectively, emptied [11].

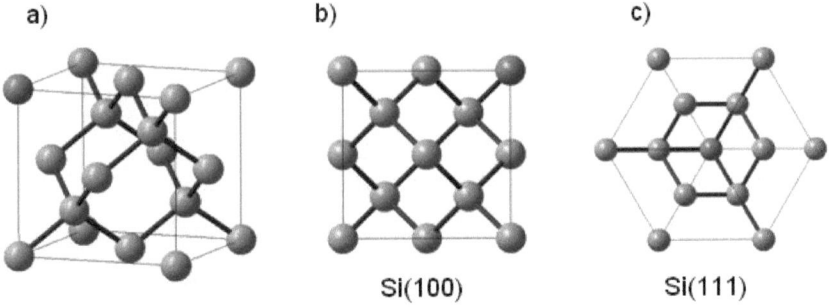

Fig. 1-1: The silicon unit cell. Cell edges are represented as thin lines. Covalent bonds are shown as thick lines. Viewing directions: (a) perspective view; (b) towards the (100) plane; (c) towards the (111) plane.

The bulk material properties of silicon depend on the crystal growth technique and further addition of electronically active impurities. Large silicon ingots can be manufactured by melt-growth according to the Czochralsky (Cz) procedure. The float-zone (FZ) technique is used to reduce the oxygen content which is typically of the order 10^{18} cm^{-3} for Cz-grown crystals. Other techniques as molecular beam epitaxy (MBE) or chemical vapor deposition (CVD) are employed for the manufacturing of thin (poly-)crystalline films [12]. In this work, silicon wafers were used manufactured by the Cz-method. FZ-material was used in control experiments whenever necessary.

Silicon dioxide (SiO$_2$) exhibits a tetrahedral structure, too (see Fig. 1-2a). Beside crystalline modifications as quartz, tridymite and crystobalite, no long-range order, however, is preserved in the amorphous dioxide and SiO$_2$ forms a network with four to eight tetrahedrons as illustrated in Fig. 1-2c. The wide and open structure of the material is due to the variation of the Si-O-Si bond angle between two adjacent tetrahedrons. This angle is nominally 145° but can cover a range between 120° and 180° (see Fig. 1-2b). Rotation around the O-axis provides an additional degree of freedom. The Si-O distance is 1.6 Å while two oxygen atoms are separated by 2.27 Å.

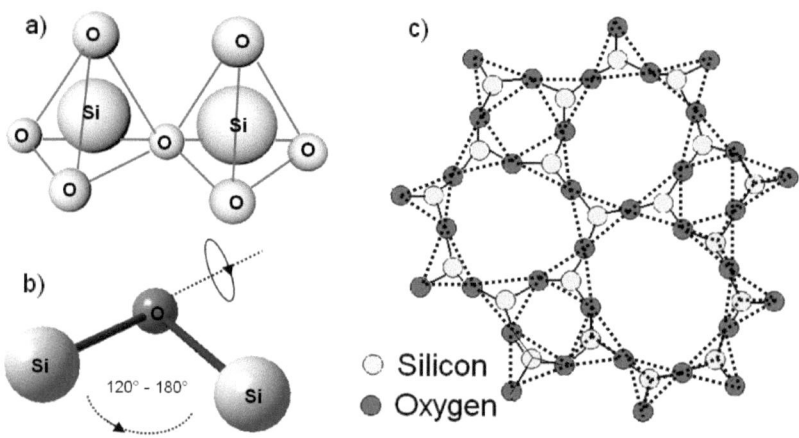

Fig. 1-2: Structural properties of silicon dioxide. (a) Tetrahedral SiO$_2$ structure. (b) Rotational and angle variations of the Si-O-Si bond structure. (c) SiO$_2$ bulk structure. Tetrahedral bonds of a silicon atom to four surrounding oxygen atoms are indicated by dashed lines. Bonds reaching out of the depicted plane were omitted.

As for bulk silicon, the properties of the dioxide strictly depend on the applied formation process and subsequent annealing steps. Thermal oxidation is, in industrial processing, the chief fabrication technique and results in high quality bulk and interface properties [13]. Wet chemical oxidation as, for instance, in a 4:1 mixture of H$_2$SO$_4$: H$_2$O$_2$ is characterized by stoichiometric variations of the oxygen coordination number (SiO_x, x = 1...4), formation of hydroxylized silicon (Si-OH) and further structural and chemical imperfections [14, 15]. Anodic oxidation of silicon in the presence of an aqueous fluoride containing electrolyte results in *wet* oxides, i.e. oxides with large Si-OH content and water inclusions [16]. As a consequence, anodic oxide layers have the highest etch rate and probably lowest density of all oxide structures mentioned above [17, 18]. Native oxide layers, finally, start to grow soon after wafer exposure to the ambient [19]. The native oxidation process results mainly from the contact of silicon with humid air [20] and can be regarded as an uncontrolled chemical oxidation process.

1.1.2 Properties of the SiO_2-Si interface

Differences in the respective densities and atomic structures necessitate the formation of a transition region at the interface of silicon dioxide and silicon. The density of silicon is 2.239 g/cm^3 while the corresponding density of SiO_2 varies between 2.0 and 2.3 g/cm^3. At the SiO_2/Si interface, these structural variations, i.e. the volume mismatch, have to be compensated by a transitional region that extends over several monolayers on both sides of the interface [21, 22]. The properties of this region are still controversially discussed in the literature for several reasons. On one hand, some of the employed experimental techniques have pronounced sensitivities to specific interface properties which are not in the scope of other methods. On the other hand, the results are markedly dependent on the oxide fabrication technique which can tremendously alter the oxide quality. Substoichiometry in the bond configuration between silicon and oxygen is now an unquestioned result that has to be regarded as the response to so-called dangling bonds at the silicon surface, i.e. hybrid orbitals which are not connected to other silicon atoms, and to the volume mismatch between SiO_2 and Si. In Fig. 1-3, a and b, two tetrahedral silicon configurations are shown for the most important surface orientations, Si(100) and Si(111). While the Si(100) surface is characterized by two dangling bonds (Fig. 1-3a), the Si(111) surface has only one bond without connection to another silicon atom (Fig. 1-3b). Model structures for an idealized interface between SiO_2 and Si are shown in the same figure [23, 24], simulating the structural transition between the two adjacent materials. The change in the oxygen coordination number, depicted in these models, is in general agreement with experimental results obtained by high-resolution photoelectron emission of the Si 2p core level shifts of ultra-thin oxide layers on Si(100) and Si(111) (Himpsel et al. [25]). The results, presented in Fig. 1-4, show additional structures next to the Si 2p main emission line (set to 0 eV) at lower binding energies which have to be attributed to the presence of varying oxidation states of silicon.

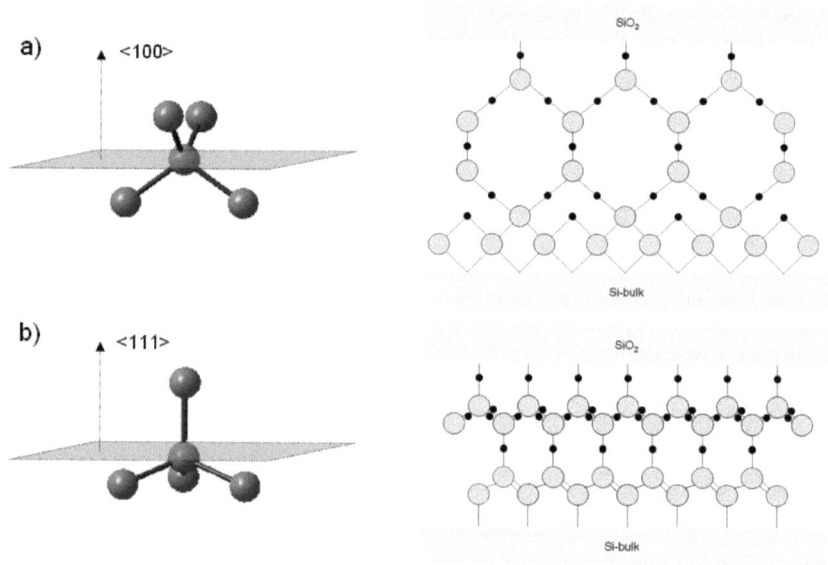

Fig. 1-3: Left: indication of the dangling bond number for the (100) and (111) orientation. Right: model interface structures between SiO_2 and Si(100) according to Yasaka et al. [23] and between SiO_2 and Si(111) according to Ohishi et al. [24].

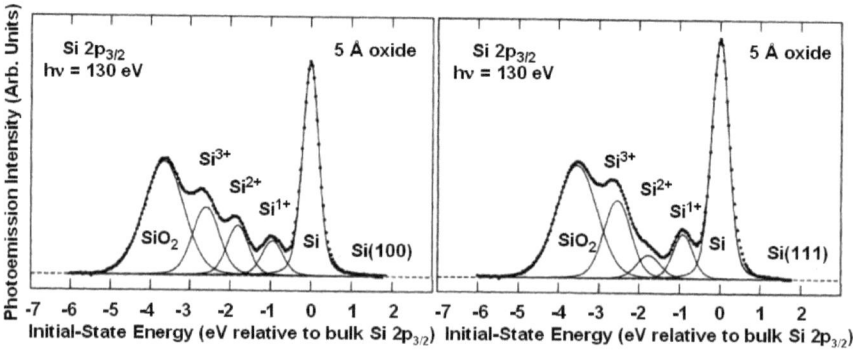

Fig. 1-4: Photoelectron emission results for the Si 2p core level shifts of Si(100), left, and Si(111), right, for ultra-thin oxide layers according to Himpsel et al. [25].

The varying oxygen coordination of a silicon atom is usually denoted as Si^{1+}, Si^{2+}, Si^{3+} and Si^{4+} although no ionic charging is present but rather partial charges of the order $\delta \sim -0.5eV$

due to the more electronegative oxygen atom. Alternatively, the different oxidation states are denoted, in stoichiometric terms, as Si_2O, SiO, Si_2O_3 and SiO_2 where the latter denotes the bulk silicon dioxide. The deconvolution of the integral XPS signal in Fig. 1-4 shows larger contributions of Si^{1+} for the (111) orientation while Si^{2+} is more pronounced at the Si(100) interface. These results make clear that structural differences are compensated by stoichiometric variation across the interface and that the interface properties also depend on the surface orientation of the silicon substrate. Many experimental results were reported that point to the existence of further transitional effects on both sides of the interface. Chemically, oxygen incorporation into the silicon bulk is assumed to stabilize the transitional region [6]; electronically, stretched bonds and non-saturated dangling bonds give rise to the formation of interface states with the typical U-shaped density distribution across the band gap [26]; structurally, distortion of the silicon atoms and bond stretching in silicon as well as in the oxide results in deformation and density variations [27]. Roughness, as a phenomenon present at the top of the silicon surface, is furthermore closely related to the density increase of interface states and is able to degrade the electronic properties of the SiO_2/Si system [28-30].

1.1.3 Stress and strain at the interface

While the incorporation of oxygen precipitates into the bulk has been extensively investigated over decades, resulting in detailed model description and reliable measurement techniques (see [31] and references therein), no unanimous agreement can be stated about structure, depth or stoichiometry of the suboxidic silicon layer. Various models are favoring either an abrupt interface consisting of a single Si^{2+} layer [32], a chemically graded system with a distribution of suboxides over a range of 2 nm [33], or a combination of oxygen incorporation in the transition layer and strained silicon bonds [34]. Strain at the silicon interface gained considerable attention by the successful exploitation of strain-induced higher electron and hole mobilities in Intel's 90 nm technology [35]. Novel nanoelectronic devices based on, e.g., strained Si-Ge or $Si-SiO_2$ heteroepitaxial interfaces are therefore subject to intensified research efforts, aiming to further miniaturization and optimized device performance [36]. Recent investigations showed that porous silicon (see sections 1.3.2 and chapter 3) is likewise applicable to induce the desired strain effects on crystalline silicon substrates [37, 38].

The theory of stress and strain often refers to a range where matter remains in its elastic range, i.e. where elongations or deformations are reversible after external forces are acting not any longer [39]. In fracture mechanics, the effect of crack formation and propagation is studied, i.e. material effects are investigated within and beyond the elasticity range [40]. Since many materials in engineering such as concrete and steel are exposed to high pressure, axial or shear forces, this field in physics is of outstanding technical importance. But also on the atomic level, the presence of stress and strain at interfaces has its impact, e.g., on the performance of devices built by hetero-junctions like the SiO_2/Si system. Unlike the investigation of macroscopic solids, the assessment of stress at buried interfaces is difficult to achieve. High-resolution X-ray diffraction patterns show a broadening of the so-called *rocking-curves* for thermally oxidized Si(100) and Si(111) [8, 41]. These results were interpreted as consequence of a strain field present at the respective silicon interfaces. In these investigations, FZ-grown silicon was used in order to exclude the influence of high oxygen concentration as observed in Cz-material.

In the elastic range of matter, strain is, according to Hooke's law, proportional to the stress force and can be derived from tensor calculations [42]. However, this description is restricted to mere mechanical effects. Other contributions which influence the stress energies at the atomic level are of chemical or electrostatic nature. According to recent calculations, applying density functional theory (DFT), the incorporation of oxygen into the silicon lattice is preferred for thermodynamic reasons. This effect influences the bond energies of silicon atoms by polarization and, chemically, by partial oxidation and contributes therefore to the total stress [6, 43]. For this reason, stress and strain at SiO_2-Si interfaces are often considered individually because of chemical and electrostatic effects which are not covered by Hooke's law.

The effect of stress forces onto the topographic structure of the SiO_2-Si(100) interface, according to a recent analysis [44], is shown in Fig. 1-5. By application of ion scattering, topographic representations of the atom displacements and bond distortions have been achieved. Scattering was measured in the channeling mode: any yield of scattering signals must result from atoms displaced from their regular position within the crystal. The analysis, however, is not unambiguous and required the solution of the inverse scattering problem, i.e. the computation of possible atom configurations near the surface that correspond to the scattering yield. In Fig. 1-5, two surface representations are shown which were in agreement to the experimental findings. The bars, given on the right side, illustrate the average silicon atom displacement throughout the first four silicon layers. The height of the bars corresponds to the standard deviation (or root-mean-square value) associated with the average displacement. According to these results, atom displacements of the range of 0.1 – 0.7 Å were calculated for a thermally oxidized Si(100) sample.

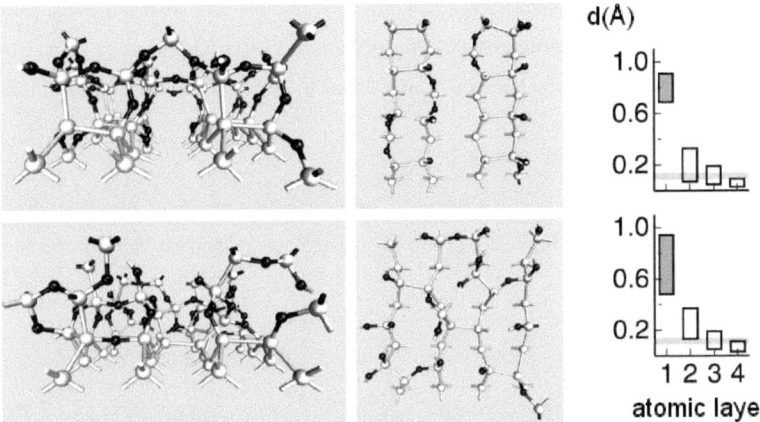

Fig. 1-5: Two possible solutions to the inverse scattering problem at a silicon surface characterized by atom displacements of the order 0.1 – 0.7 Å. Left: arrangement of distorted Si atoms. Right: average silicon atom displacements are shown for the first four silicon layers. The height of the rectangular bars represents the root-mean-square value of the average atom displacement. Images and data were taken from [44].

1.2 Competing electronic and (photo-)electrochemical processes at the reactive semiconductor-electrolyte interface

The electron transfer in homogeneous solutions, according to the theory of R. A. Marcus [45, 46], is considered first. The reorganization energy of the solvation-shell / ion system is thereby introduced. The distribution of energy states at the semiconductor-redox interface, based on the work of H. Gerischer [47-49], will be subsequently presented together with the corresponding description of energy levels in semiconductors.

1.2.1 The Marcus theory of single electron transfer

Ions in solution of a polar solvent are surrounded by oriented and polarized solvent molecules. Due to thermal motion, the energy of the ion-solvent system is subject to fluctuations which can be regarded, in first approximation, as the corresponding energy of a harmonic oscillator. Any change of the electronic configuration of the central ion will affect, by electrostatic interaction, the positions of the solvent molecules, i.e., a new equilibrium configuration requires rearrangement and reorganization of the molecules. According to the *Frank-Condon principle* [50], the electron transition probability is highest when the energies of the electron in the initial and final states are identical. From this principle follows that the electron transfer takes place prior to any rearrangement. This phenomenon was addressed by R. A. Marcus by a model which considers the transfer from a donor molecule, D, to an acceptor molecule A, i.e. the reaction:

$$D + A \rightarrow D^+ + A^-. \tag{1-1}$$

The energetics of the reaction is illustrated within the framework of the generalized phase space where the Gibbs free energy of the system is related to the so-called reaction coordinate(s) q.

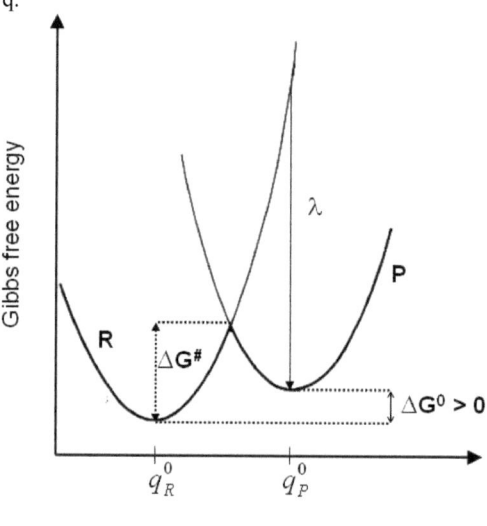

Reaction coordinate q

Fig. 1-6: Gibbs free energy curve(s) for the electron transfer between a donor molecule D and an acceptor molecule A. The pairs R = (D, A) and P = (D$^+$, A$^-$) represent the reactant and the product, respectively. The Gibbs energy of activation is indicated as $\Delta G^{\#}$, the reorganization energy as λ. The situation is shown for an endergonic reaction, i.e., $\Delta G^0 > 0$ (according to [51]).

Here, the variable q represents a suitable choice of independent variables, derived from all nuclei positions of the solvent-molecule-system including the mutual distance between D and A. A trajectory in the phase space, as illustrated in Fig. 1-6, represents then all possible states of the pair R(D,A), describing the reactant, and P=(D$^+$,A$^-$), describing the product. The equilibrium positions of the reactant and the product are indicated as q_R^0 and q_P^0, respectively. All other points along the curves represent intermediate states far from equilibrium. It should be noted that the parabolic curves result from the treatment of the system as harmonic oscillator. This visualization simplifies the real situation where multi-dimensional energy surfaces have to be considered rather than two-dimensional curves. As shown in Fig. 1-6, the electron transfer requires usually some activation energy $\Delta G^{\#}$. Subsequent to the electron transfer, the reorganization of the solvent molecules allows the product P to reach its equilibrium position by release of the energy λ, the reorganization energy. Depending on the energy differences of R and P at equilibrium, endergonic, isoergonic and exergonic processes can be distinguished with $\Delta G^0 > 0$, $\Delta G^0 = 0$ and $\Delta G^0 < 0$, respectively, i.e. reactions that consume or release energy. Using the parabolic-curve representation for transfer reactions, also activation-free (R_2) and barrierless (R_3) reactions can be described as shown in Fig. 1-7. For an activation-free reaction $\Delta G^{\#} = 0$ holds while the barrierless reaction requires no reorganization subsequent to the electron transfer.

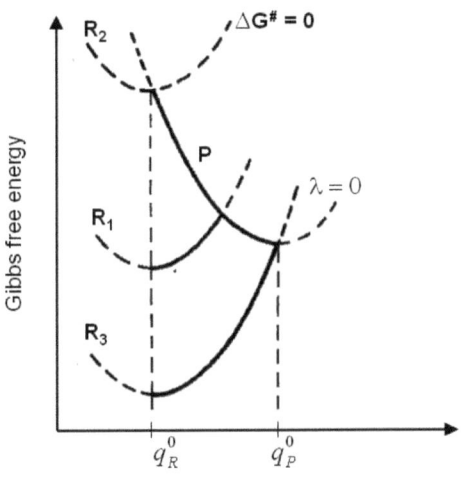

Fig. 1-7: Gibbs free energy curves for an activation-free reaction (R_2) and a barrierless reaction (R_3). The curve R_1 corresponds to the situation in Fig. 1-6.

If, for a given system, the electron transfer probability, the kinetics of solvent fluctuation and the energetics are known quantities, then an electron transfer rate can be derived [52]:

$$k_{el} = \kappa \nu \exp\left(-\frac{(\Delta G^0 + \lambda)^2}{4kT\lambda}\right). \qquad (1\text{-}2)$$

Here, κ denotes the electron transmission coefficient which expresses the probability of the electron transfer, ν denotes the frequency of nuclear motion in terms of the harmonic oscillator, ΔG^0 describes the change in Gibbs free energy and λ the reorganization energy.

So far, no interaction between the reacting molecules was assumed. However, the molecules have to approach each other very closely to reach a tunneling distance over which the electron can be transferred. Coupling of the respective electron systems is then expected which results in splitting of the energies near the point of intersection (see Fig. 1-8).

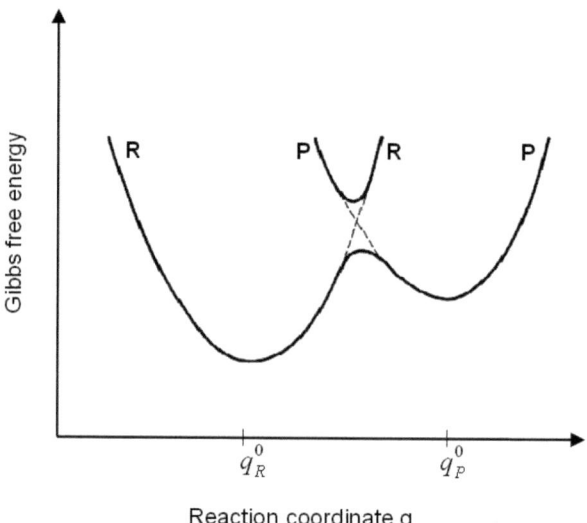

Fig. 1-8: Splitting of the energy curve due to electron coupling for reactants being very close to each other. Both curves are well separated by the gap which corresponds to the adiabatic case considered in the Marcus theory.

Electronic coupling of the electronic states can be expressed in quantum mechanical terms as:

$$V_{RP} = <\psi_R^0 |H| \psi_P^0 > \qquad (1\text{-}3)$$

where H denotes the total electronic Hamiltonian while ψ_R^0 and ψ_P^0 represent the respective equilibrium wave functions of the reactant and the product [52]. The gap between the curves in Fig. 1-8 depends on the magnitude of V_{RP}. For large values of V_{RP}, the curves are well separated and the reaction pathway follows the lower energy curve in Fig. 1-8. For small values, however, which are characteristic for non-adiabatic reactions, the point of intersection is only slightly disturbed. In this case, the electron transfer is less probable and the reactant moves along parabola R without electron transfer. In other words: if the time for reorganization of the solvent molecules is large compared to the process of electron transfer, then electron transfer is more probable. This description corresponds to a slow movement along the respective parabolas in Figs. 1-6 through 1-8 and coincides with the physical fact that reorganization of the solvent molecules requires much more time (10^{-13} to 10^{-11} s) than the electron tunneling process (about 10^{-15} s) [52]. In this case, the transmission factor κ in the rate equation (Eq. 1-2) is close to unity.

1.2.2 Energy levels in semiconductors and redox systems

In the preceding section, energy levels of solvated molecules/ions in a homogeneous solution were considered. The corresponding energy levels in semiconductors require information about the band structure of the material [9, 11]. Conduction and valence bands of a semiconductor (E_c and E_v, respectively) are separated by an energy gap, E_g. The Fermi energy E_F, for a non-degenerated semiconductor, is located within this interval, some kT off the respective band edges where k denotes the Boltzmann constant and T the temperature. With the Fermi distribution function, f(E), and the density of states, N(E), the charge carrier concentration within the respective energy bands can be calculated from:

$$n = \int_{E_C}^{\infty} f(E)N(E)dE \quad \text{and} \tag{1-4}$$

$$p = \int_{-\infty}^{E_V} (1-f(E))N(E)dE . \tag{1-5}$$

For low temperatures and a non-degenerated semiconductor, the respective densities of states can be approximated by the corresponding density of the free electron gas:

$$N_C = N_C^{eff}(E-E_C)^{1/2} = \frac{8\sqrt{2\pi}}{h^3}(m_e^*)^{3/2}(E-E_C)^{1/2} \quad \text{and} \tag{1-6}$$

$$N_V = N_V^{eff}(E-E_V)^{1/2} = \frac{8\sqrt{2\pi}}{h^3}(m_h^*)^{3/2}(E_V-E)^{1/2} . \tag{1-7}$$

Here, $m_{e/h}^*$ denotes the effective electron/hole mass. For silicon, the effective densities of states, $N_{C/V}^{eff}$, are $N_C^{eff} = 2.8 \times 10^{19}$ cm^{-3} and $N_V^{eff} = 1.04 \times 10^{19}$ cm^{-3}. This approximation yields:

$$n = N_C^{eff} \exp\left(-\frac{E_C-E_{F,n}}{kT}\right) \quad \text{and} \tag{1-8}$$

$$p = N_V^{eff} \exp\left(-\frac{E_{F,p}-E_V}{kT}\right) \tag{1-9}$$

for the concentrations of electrons and holes, respectively.

At equilibrium, i.e. for $E_F = E_{F,n} = E_{F,p}$, and $n = n_0$, $p = p_0$, the product

$$n_0 p_0 = N_C N_V \exp\left(\frac{E_C - E_V}{kT}\right) = n_i^2 = const. \qquad (1\text{-}10)$$

is determined by the intrinsic carrier concentration n_i of silicon which is about $1.45 \times 10^{10} cm^{-3}$ at room temperature. The equilibrium carrier concentrations n_0 and p_0 in the conduction and valence band can be calculated from Eqs. 1-8 and 1-9. The relation 1-10 allows, e.g., the calculation of the minority carrier concentration if the doping density and therefore the majority carrier concentration is known.

The semiconductor Fermi level corresponds to the electrochemical potential of electrons in a redox system that can, according to the Nernst equation [53], be described by:

$$\mu_{e,redox} = \mu_{redox}^0 + kT \ln\left(\frac{c_{ox}}{c_{red}}\right). \qquad (1\text{-}11)$$

Here, the concentrations of the oxidized and reduced species in solution are denoted as c_{ox} and c_{red}, respectively, while μ_{redox}^0 depends on the applied reference system. In solid state physics, this reference is given by the vacuum level of the absolute energy scale. In electrochemistry, reference energies are obtained by use of the normal hydrogen or saturated calomel electrode (NHE or SCE, respectively). In this case, the Fermi level of the redox system coincides with the electrochemical potential, i.e.

$$E_{F,redox} = \mu_{e,redox}. \qquad (1\text{-}12)$$

In order to refer both scales to each other, the work function of, e.g., the normal hydrogen electrode has to be measured. Then, the relationship of the semiconductor Fermi level, on the absolute scale, with the redox potential of an electrolyte, measured against NHE, can be expressed as:

$$E_{F,redox} = E_{ref} - eU_{redox} = E_{F,redox} = -4.5 \ eV - eU_{redox}. \qquad (1\text{-}13)$$

On the right-hand side, an average value for the work function of the NHE of -4.5 eV was chosen. It should be noted that NHE work function values, obtained by various experimental investigations and theoretical calculations, differ by about ± 0.2 eV [54, 55].

In the Gerischer model [49], the densities of energy states in a redox-electrolyte can be described by the electronic energy levels of the oxidized and reduced species in solution and by the reorganization energy associated with the single electron charge transfer process. These quantities contribute as parameters to the distribution of energy levels of the oxidized (D_{ox}) and reduced species (D_{red}) in form of a Gaussian normal distribution:

$$D_{ox} = \frac{1}{\sqrt{4kT\lambda}} \exp\left(-\frac{(E - E_{F,redox} - \lambda)^2}{4kT\lambda}\right) \text{ and} \qquad (1\text{-}14)$$

$$D_{red} = \frac{1}{\sqrt{4kT\lambda}} \exp\left(-\frac{\left(E - E_{F,redox} + \lambda\right)^2}{4kT\lambda}\right). \qquad (1\text{-}15)$$

The contact formation between the semiconductor and the redox-system is achieved by charge transfer across the interface until the respective Fermi levels are equal, i.e.

$$E_F = E_{F,redox}. \qquad (1\text{-}16)$$

Usually, the charge carrier density of the electrolyte is much higher than the corresponding density of the semiconductor and the potential drop occurs across the semiconductor rather than the electrolyte. In the case of complete carrier depletion in the semiconductor, the width, d_{sc}, of band bending can be calculated by the *Mott-Schottky approximation*:

$$d_{sc} = \sqrt{\frac{2\varepsilon_0\varepsilon}{eN_D}\left(\Phi_{sc} - U - \frac{kT}{e}\right)}. \qquad (1\text{-}17)$$

Eq. 1-17 refers to an n-type semiconductor of doping density N_D. Φ_{sc} denotes the potential drop across the space-charge region (SC) with respect to the potential in the bulk and U denotes any externally applied potential.

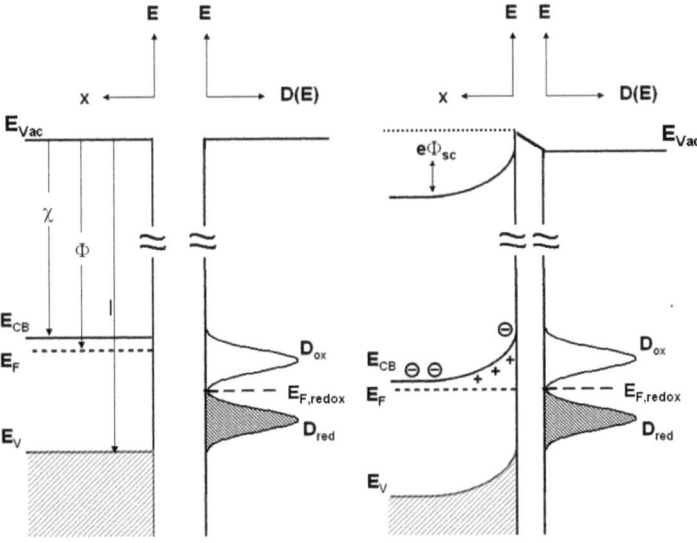

Fig. 1-9: Semiconductor and redox system before (left) and after contact (right). Due to the higher carrier concentration of the liquid, band banding occurs almost exclusively in the semiconductor. The drop of the potential across the Helmholtz layer is also indicated.

In order to describe the contact formation between the redox-system and the semiconductor, the presence of ionic charges at the interface has to be considered which form a double layer, the so-called Helmholtz double-layer, which results in an additional potential drop. If no further charges have to be considered such as electronically active surface states, dipole layers or chemically adsorbed species, the contact formation can be illustrated as in Fig. 1-9. Here, the electron affinity, χ, the work function, Φ, and the ionization energy I, are indicated as important for the characterization of the semiconductor. The effect of the Helmholtz-double-layer, built by two semi-layers of opposite charge, is shown in Fig. 1-10. The inner Helmholtz-layer, IHL, consists of ions which have lost parts of their solvation shells (specific adsorption). The outer Helmholtz-layer, OHL, is built by solvated ions of opposite charge.

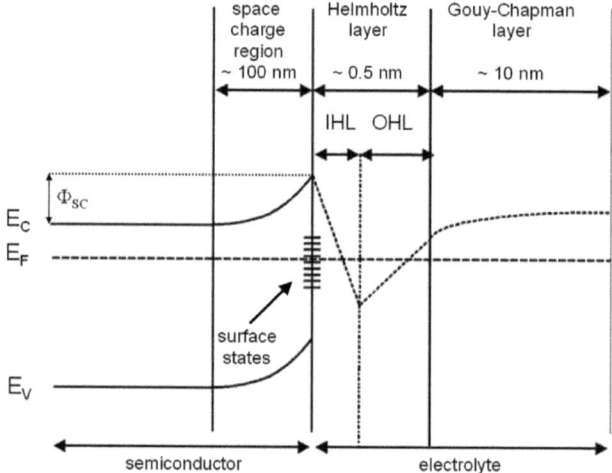

Fig. 1-10: N-type semiconductor in contact with a redox system: the potential distribution varies across the space charge region, the Helmholtz-double-layer and the Gouy-Chapman layer. The electrochemical potential can be additionally influenced by surface states.

Additionally, the Gouy-Chapman layer is indicated which extends into the electrolyte bulk for lower electrolyte concentrations and results in a locally varying potential drop. For experiments, to be described in section 3.2, the electrolyte concentration of the used 0.1 M NH_4F solution is high enough such that consideration of the Gouy-Chapman layer may not be required. In concentrated NH_4F (40%), as employed for silicon microstructuring in section 3.3, the potential drop occurs only across the Helmholtz-double-layer. Furthermore, surface states are shown which influence the equilibrium position of the electrochemical potential.

In accordance to the Marcus theory, described above, the sequential steps of charge transfer and solvent reorganization at the semiconductor-electrolyte interface can be described by the Gerischer model [47, 48]:

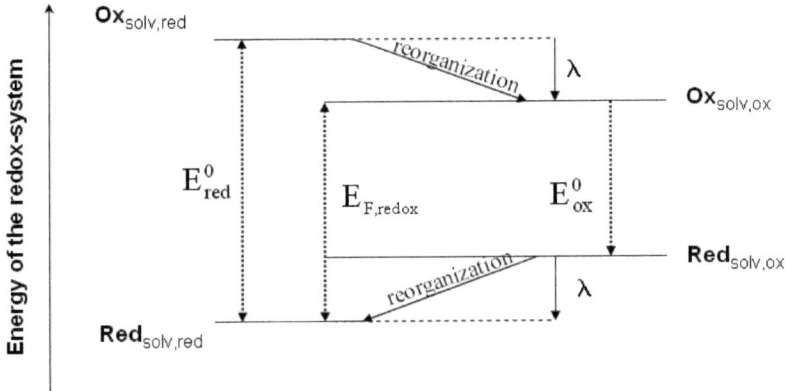

Fig. 1-11: Charge transfer between a reduced ion-solvent and an oxidized ion-solvent system. The reorganization energy λ is released and, respectively, absorbed to complete the transition between Red$_{solv,red}$ and Ox$_{solv,ox}$. Ox$_{solv,red}$ (Red$_{solv,ox}$) describe the species after (before) electron transfer but before reorganization of the solvation shell.

An energy E_{red}^0 is required to bring an electron of the reduced species, R$_{solv,red}$, to the vacuum level. The electron transfer is fast such that the solvation shell remains in its initial configuration (Frank-Condon principle). The following reorganization of the solvation shell lowers the energetic level of the oxidized molecule by the reorganization energy λ. Analogously, the electron transfer from the vacuum level to the oxidized molecule lowers its energetic position by E_{ox}^0 before reorganization of the solvation shell takes place. The energetic difference between the oxidized and reduced species at their respective equilibrium positions (Ox$_{solv,ox}$ and Red$_{solv,red}$) determines the redox potential $E_{F,redox}$ of the reaction.

1.2.3 Semiconductor (photo-)corrosion

Decomposition of a semiconductor electrode requires charge transfer reactions to or from the respective energy bands (conduction or valance band) or surface states. Under anodic potential, electron injection into the conduction band of an n-type semiconductor in the dark or hole extraction from the valence band under illumination can initiate the electrochemical modification of the electrode. While corrosive processes in the dark can be understood by consideration of the semiconductor Fermi level with respect to the energetic position of the potential in the electrolyte, corresponding reactions under illumination are determined by the individual Fermi levels of electrons and holes. These quasi-Fermi levels, $E_{F,n}$ and $E_{F,p}$, can be calculated according to:

$$E_{F,n} = E_F + kT\ln(1+\frac{\Delta n(z)}{n_0}) \text{ and} \qquad (1\text{-}18)$$

$$E_{F,p} = E_F - kT\ln(1+\frac{\Delta p(z)}{p_0}), \qquad (1\text{-}19)$$

where n_0 (p_0) denote the carrier concentrations in the dark, $\Delta n(z)$ ($\Delta p(z)$) the corresponding concentrations under illumination. The dependence of the carrier concentration on z, which points in the direction of light propagation, considers the decreasing concentration of photogenerated carriers with increasing penetration depth. The penetration depth, in turn, is dependent on the absorption coefficient α of the material (see section 2.1.1). If the incident photon flux, Φ_0, is known, the concentration of light induced carriers is determined by:

$$g = \alpha\Phi_0 \exp(-\alpha z). \qquad (1\text{-}20)$$

The corresponding band diagram of an illuminated n-type semiconductor is illustrated in Fig. 1-12:

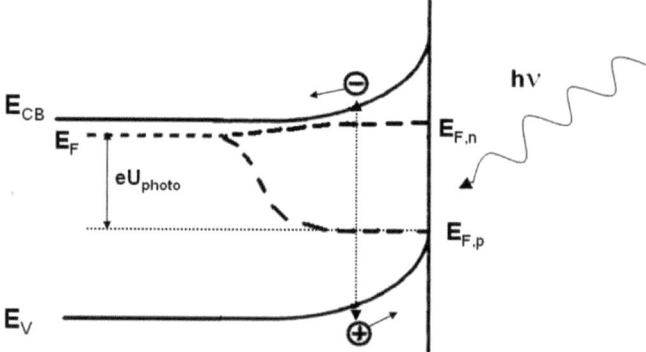

Fig. 1-12: Splitting of the Fermi energy level into two quasi-Fermi level energies $E_{F,n}$ and $E_{F,p}$ for electrons and holes, respectively, upon illumination by light of photon energy hv. The photovoltage U_{photo} is determined by $(E_{F,p} - E_F)/e$.

The analogous scheme for a p-type electrode has to consider the different position of the Fermi-level, close to the valence band, and the reverse charge transfer reactions under cathodic potential, i.e., electron extraction from the conduction band and, respectively, hole injection into the valence band. For anodic potentials, no photogenerated charge carriers are required.

In the present work, (photo-)corrosive processes on silicon, immersed in ammonium fluoride (NH₄F) containing solutions are of primary interest. The dissolution mechanism is therefore reviewed for both the chemical and electrochemical dissolution. Model considerations of the chemical dissolution process usually consider a Si(111) surface whose dangling bonds are saturated by hydrogen atoms (H-termination). This termination is effective in suppressing the instantaneous oxidation of silicon surface atoms in contact with ambient air [56]. The dissolution reaction in aqueous NH₄F solution is essentially determined by the exchange reaction of Si-H for Si-OH and finally Si-F as illustrated by the following scheme, proposed by P. Allongue, V. Kieling and H. Gerischer [57]:

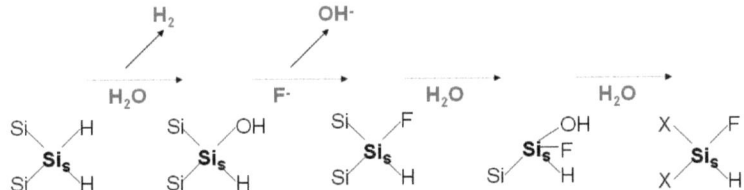

Fig. 1-13: Chemical dissolution of Si(111) in NH₄F containing solutions. The process is described for a surface atom, S$_S$, at a kink site, i.e., an atom with only two backbonds. Arrows indicate the consumption of water (H₂O) and fluorine ions (F⁻) and, respectively, the release of hydrogen (H₂) and hydroxyl ions (OH⁻) during the sequence of chemical steps.

In Fig. 1-13, the dissolution of a silicon surface atom, S$_S$, with only two backbonds (a kink site atom) is considered. It is assumed that the dissolution of silicon takes place at the highest rate for these kink site atoms while fully coordinated atoms on flat terrace sites with three backbonds exhibit a dissolution rate which is, e.g. in 40% NH₄F, lower by a factor of 10^7 [58]. The scheme describes the removal of a surface atom by attack of a water molecule and subsequent exchange of OH for F. The atom is finally released into the solution where possibly full hydrolysis takes place.

For the photoelectrochemical dissolution of silicon, the overall reactions can be described by the following equations:

(divalent dissolution) $Si + 6HF + h_{VB}^+(h\nu) \rightarrow SiF_6^{2-} + 4H^+ + H_2 + e_{CB}^-$ (1-21)

and

(tetravalent dissolution) $Si + 2H_2O + 4h_{VB}^+(h\nu) \rightarrow SiO_2 + 4H^+$. (1-22)

In Eqs. 1-21 and 1-22, the photogenerated holes, required for the reaction, are indicated as $h_{VB}^+(h\nu)$ in order to refer to the photon energy of the incident light and the valence band where the holes are originating. The divalent dissolution proceeds by extraction of a valence band hole and injection of an electron into the conduction band. Subsequently, the silicon atom is dissolved into the solution. The tetravalent dissolution, in turn, oxidizes the silicon surface atoms, i.e. an oxide film is left behind. Depending on the pH of the solution, this oxide is dissolved by the two reactions:

$$SiO_2 + 6HF \rightarrow SiF_6^{2-} + 2H_2O + 2H^+ \qquad (1\text{-}23)$$

(for pH < 4) and

$$SiO_2 + 3HF_2^- \rightarrow SiF_6^{2-} + H_2O + OH^- \qquad (1\text{-}24)$$

(for 2 < pH < 4).

Oxide etching according to reaction 1-24 is faster by a factor of about 4 than 1-23.

In neutral or alkaline solutions, F⁻ is the predominant species in solution, i.e., HF molecules are completely dissociated. Fewer details are known about oxide dissolution in 40% NH₄F. Only recently, a study comprising photoelectron spectroscopy, ellipsometry and atomic force microscopy was published indicating that 40% NH₄F etches silicon dioxide at all [59]. The publication, however, does not suggest reaction pathways but it can be assumed that, in principle, the same steps apply as for lower pH: the SiO₂ bonds are attacked by fluorine ions, the oxygen atom transforms to H₂O by protonization and SiF₄ reacts to SiF_6^{2-} in the solution.

1.3 Self-organization phenomena at the silicon/electrolyte interface

1.3.1 Dynamical systems

A dynamical system [60, 61], according to its abstract mathematic definition, is any equation or set of equations that can be described as

$$\dot{x}(t) = F(x,t). \qquad (1\text{-}25)$$

Here, x denotes the dependence on the coordinates in space and the set of parameters that describes the full system (the phase space) while t denotes the corresponding dependence on time. The system may or may not be explicitly dependent on time as suggested by the right hand side and nothing is said about the properties of F. In many cases F exhibits pronounced nonlinear behavior [62]. In either case, solutions to Eq. 1-25 can be illustrated by a trajectory in the phase space which is built by the set of all (generalized) coordinates of the system. Particular properties of dynamical systems, as oscillatory or convergent behavior, are described to the very detail in the literature [63]. Many of these properties result from intricate nonlinearities which impede both analytic solution of the corresponding equations and long-term prediction. As a consequence, mathematical treatment of these systems focuses on classificatory terms such as stability, convergence or oscillatory behavior rather than detailed description of the trajectories.

1.3.2 Self-organization phenomena at silicon electrodes

Self-organization in electrochemical dynamical systems can be generally described as the phenomenon of spontaneous temporal or spatial pattern formation [64]. From a mathematical point of view, sets of differential equations are adequate to describe a system that sensitively responds to the variation of its possible states. Particularly, nonlinearities in this set of equations are considered to invoke spontaneous pattern formation as soon as some parameters reach specific magnitudes. Instabilities in the system behavior may then arise and result in the observed formation of recurrent patterns. The time- or space-scale on which self-organization is observed, usually exceeds considerably the time-scale on which the reaction kinetics takes place or the space-scale of the atomic, molecular or ionic constituents of the system. Self-organization therefore means the formation of extended patterns in time or space. This explanation is based on the criterion that regularity is recognizable *after* the patterns have formed. But self-organization has also an important further aspect *during* the patterns are forming: the system is able to respond to the variation of external parameters by a so-called negative feedback. This type of system-response preserves the patterns in their characteristic form although experimental conditions impose some perturbations. In electrochemical systems, e.g., pattern evolution may be preserved while potential, solution concentration or pH value are varied within a certain range. Mainly focusing on metal electrodes, the intricate mathematical treatment of self-organization is extensively described in [64].

Examples of self-organization during electrochemical conditioning of silicon extend from the oscillatory behavior in time to the large variety of pore formation in space. Photocurrent oscillations in diluted HF solutions were first described by Turner in 1958 [65]. Since then, tremendous efforts were made in order to understand the finding and to predict the behavior in dependence on the electrochemical conditions. Lehmann, e.g., assumed varying etch rates during the oscillation process [66] while Lewerenz [67] favored constant etch but varying oxidation rates; a statistical treatment of the period of local oxide oscillators is an important quantity in the theoretical model of Grzanna, Jungblut and Lewerenz [68] while Chazalviel et al. and Carstensen et al. emphasize synchronization phenomena between locally oscillating domains [69, 70].

In Fig. 1-14, photocurrent oscillations on n-type silicon are shown. The assumption of local oxide oscillators, characterized by varying growth rate and oscillation periods, led to a model that almost perfectly imitates the global coupling as measured by the integrally oscillating photocurrent (squares in Fig. 1-14) [71].

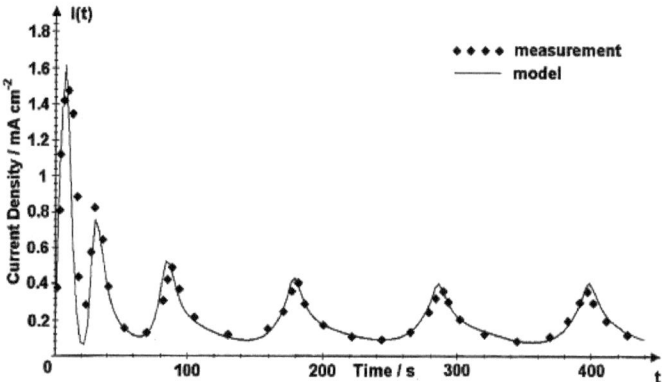

Fig. 1-14: Photocurrent oscillations, measured for n-type silicon under illumination (straight line). The model, developed by Grzanna, Jungblut and Lewerenz [71], simulates the initial behavior as well as sustained oscillations at later cycles (black squares).

Spatial examples of self-organization are two-fold in the case of silicon. In 1956, the formation of microporous silicon was reported first by Uhlir [72]. The interest in this material boosted when Canham reported visible photoluminescence from porous silicon (PS) layers at room temperature in 1990 [73]. Above band-gap photoluminescence (PL) at 4.2° K from PS was observed first in 1984 [74, 75]. With the discovery of visible PL, quantum confinement was discussed to explain the observed shift of the PL to higher energies [76]. In the following years the influence of doping concentration, porosity, pore geometries and sample aging (i.e. pore wall termination by silicon dioxide) on the PL intensity was extensively investigated and led to innumerable publications.

While there is an ongoing search for higher PL efficiencies, the application of PS meanwhile exceeds the field of optoelectronics. Specific areas in which PS is being implemented comprise sensors, mass spectrometry, nanocrystal production, drug delivery, biomaterials, fuel cells and photovoltaics (see [77] and references therein). One of the most fascinating properties of PS is the formation process itself: the self-adjustment and self-limiting of the pore evolution makes this phenomenon a primary subject for studies of self-organization effects. Microporous silicon (pore diameter 2-10 nm) can be regarded as the random counterpart of self-organized patterns, revealing its inherent regularity not until some mathematical treatment such as, e.g., fractal analysis [78-81] or description in terms of Diffusion Limited Aggregation Theory (DLA) [82]. Both concepts will be addressed in more detail in section 3.2 and 3.3., respectively, when discussing the phenomenon of microporous layers beneath silicon nanocrystals and the formation of fractal-like lateral macropores.

The second example, silicon macropores (pore diameter 100 nm -20 µm), reveals its regularity at first glance. In Fig. 1-15, two examples are shown, published by V. Lehmann [83]. Macropores, visible in Fig. 1-15, left, were forming by self-organization on low-doped n-Si(100) in diluted HF solution. The application of standard lithography (see Fig. 1-15, right) restrained the pore position to pre-determined nucleation sites.

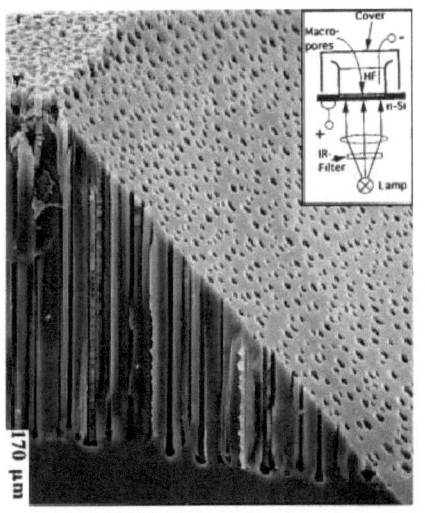

Fig. 1-15: Macropore formation on a Si(100) bevel by anodization in solutions of diluted HF. Left: randomly distributed macropores obtained after a sequence of dissolution at 10 V for 1 min and at 3 V for 149 min in 6% HF. Right: standard lithography was used to initiate the pore growth at specific sites (see inset). For subsequent electrochemical dissolution, 2.5% HF was used. Images were taken from [83].

In this research field, too, phenomenological description still outweighs the theoretical understanding and many publications are available which relate pore formation principles to varied experimental conditions such as electrolyte composition and wafer orientation [84-86]. Presently, the so-called *current-burst model* is a successful qualitative description of the findings [87] as described in more detail in section 3.3.6. In chapter 3 of this work, two further self-organization phenomena will be described. Firstly, the formation of silicon nanocrystals, protruding out of the surface, is investigated. Similarly to the structural properties of microporous silicon, mathematical filtering reveals hidden topographical regularities in dependence on small variations of the crystal wafer orientation, i.e. the miscut-angle. Secondly, lateral formation of microstructures will be analyzed which form with varying symmetry on Si(111) and Si(100) substrates. In some respect, these structures can be regarded as the lateral complement to vertical macropores, shown above.

2. Experimental methods and procedures

2.1 Brewster-angle analysis

The response of a solid to the presence of an external electromagnetic field is given by its resulting polarization [88]. This effect is essential for the understanding of optical measurements on metals, semiconductors and insulators. In Maxwell's equations, the polarization is related to the dielectric function of a solid from which the reflectance coefficients can be derived [89]. This leads to the well-known Fresnel equations. The interaction between an electromagnetic field and matter will be presented first. The experimental setup of Brewster-angle analysis (BAA) will be described afterwards, followed by the mathematical description of multi-layer analysis of smooth and rough surfaces and interfaces.

2.1.1 The dielectric function

The polarizability of matter results from the large number of atomic dipole moments that arrange under the influence of an external electric field. The linear superposition of the resulting dipole field and the external field can then be expressed as

$$\vec{D} = \varepsilon_0 \vec{E} + \vec{P} \quad , \tag{2-1}$$

where $\vec{D}$ denotes the dielectric displacement, $\vec{E}$ the electric field, $\vec{P}$ the polarization and ε_0 the dielectric constant of vacuum. Since the polarization $\vec{P}$ represents the response of matter to an external field, it can be described as a power series of $\vec{E}$. In case of spatial isotropy, polarization and electric field point towards the same direction. By introduction of the electric susceptibility χ_e, it is possible to express the polarization in dependence on the electric field by $\vec{P} = \chi_e \varepsilon_0 \vec{E}$ where terms of higher order were omitted (linear approximation). In the more general case, polarization and electric field do not necessarily coincide in their directions and the susceptibility has to be described by a tensor of rank two, i.e. by a 3 x 3 matrix. Moreover, high electric field strength requires the consideration of further, nonlinear terms which is usually subject in the field of nonlinear optics. By the definition $\varepsilon_r = 1 + \chi_e$, the relative dielectric constant is introduced and Eq. 2-1 for linear media transforms to:

$$\vec{D} = (1 + \chi_e)\varepsilon_0 \vec{E} = \varepsilon_r \varepsilon_0 \vec{E} . \tag{2-2}$$

It should be noted that ε_r is frequency-, i.e., energy-dependent. Furthermore, for absorbing matter $\varepsilon_r \equiv \varepsilon_r(\omega)$ is a complex number. The dielectric constant reads then $\varepsilon_r = \varepsilon_1 + i\varepsilon_2$. Real part and imaginary part can each be related to relevant physical quantities (see below). The relation of the dielectric constant (or function) to those phenomena known from physical experiments as light absorption or change of speed of light in its passage through a medium can only be understood by consideration of the solution of the wave equation. The travel of light in non-conductive media is determined by the homogenous wave equation as derived from Maxwell's set of equations:

$$\left(\Delta - \frac{1}{u^2}\frac{\partial^2}{\partial t^2}\right)\vec{E}(\vec{r},t) = 0 . \tag{2-3}$$

The speed of light in the material, u, is related to c, the speed of light in vacuum, divided by n, the (real) index of refraction, the dielectric constant of vacuum, ε_0, the dielectric constant of the material, ε_r, the permeability of vacuum, μ_0, and the corresponding permeability of the material, μ_r:

$$u = \frac{1}{\sqrt{\varepsilon_r \varepsilon_0 \mu_r \mu_0}} = \frac{c}{\sqrt{\varepsilon_r \mu_r}} = \frac{c}{n} . \tag{2-4}$$

In conductive media, the additional term $\vec{j} = \sigma \vec{E}$ has to be considered, relating the current density $\vec{j}$ to the electric field $\vec{E}$ by the conductance σ of the material. Applying the Drude-Sommerfeld approximation for conductive media [90], the wave equation reads:

$$\left[(\Delta - \frac{1}{u^2}\frac{\partial^2}{\partial t^2}) - \mu_r \mu_0 \sigma \frac{\partial}{\partial t}\right] \vec{E}(\vec{r},t) = 0. \qquad \text{(Telegraph equation)} \qquad (2\text{-}5)$$

Using

$$k = \frac{1}{2}n^2\left[-1 + \sqrt{1 + (\frac{\sigma}{\varepsilon_r \varepsilon_0 \omega})}\right], \qquad (2\text{-}6)$$

a solution for Eq. 2-5 can be obtained by a plane wave traveling along the z-direction:

$$\vec{E}(z,t) = \vec{E}_0 \cdot \exp\left[-(k\omega/c)z\right]\exp\left[i\omega[(n/c)z - t]\right]. \qquad (2\text{-}7)$$

The preceding exponential factor describes the damping of the wave during its travel through the material. The loss of intensity is due to transformation of wave energy in heat energy which can be considered as equivalence to the formation of heat in conductive media through which an electric current is flowing. Consequently, the quantity k is named the extinction coefficient which determines the depth of penetration of the wave.

Since light intensity is proportional to the square of the electric field, i.e., $I \propto \vec{E}\vec{E}^* = |\vec{E}_0|^2 \exp[(-2k\omega/c)z]$, the intensity decays to its e-th fraction after

$$\frac{c}{2k\omega} = \frac{\lambda}{4\pi k}. \qquad \text{(depth of penetration)} \qquad (2\text{-}8)$$

The inverse quantity

$$\alpha = \frac{4\pi k}{\lambda} \qquad (2\text{-}9)$$

defines the absorption coefficient.

In order to define, in analogy to non-conductive media, a (complex) index of refraction, the definition $N \equiv n + ik$ can be used. The (complex) dielectric constant ε_r and N can then be related to each other using Eq. 2-6

$$\varepsilon \equiv \varepsilon_r = \frac{c^2}{u^2}\mu_r + \frac{\mu_0 \sigma c^2}{\omega} = \varepsilon_1 + i\frac{\sigma}{\varepsilon_0 \omega} = \varepsilon_1 + i\varepsilon_2. \qquad (2\text{-}10)$$

It should be noted that above the definition $N \equiv n + ik$ was used for the complex index of refraction while some authors agree in using $N \equiv n - ik$. The first form leads to $\varepsilon \equiv \varepsilon_1 + i\varepsilon_2$ while the latter results in $\varepsilon \equiv \varepsilon_1 - i\varepsilon_2$ [91]. The relation between the corresponding quantities, index of refraction, n, and extinction coefficient, k, on one hand, and the components of the dielectric function, ε_1 and ε_2, on the other hand, are summarized in the following table:

n	$\mathrm{Re}(\sqrt{\varepsilon})$
k	$\mathrm{Im}(\sqrt{\varepsilon})$
$\mathrm{Re}(\varepsilon) = \varepsilon_1$	$n^2 - k^2$
$\mathrm{Im}(\varepsilon) = \varepsilon_2$	$2nk$

Table 2-1: Relation of index of refraction, n, extinction coefficient, k, to real and imaginary part of the complex dielectric function $\varepsilon \equiv \varepsilon_1 + i\varepsilon_2$.

The physical meaning of ε_2 for optical measurements can be derived from quantum mechanical treatment of the interaction of a solid with an electromagnetic field. Using perturbation theory of first order for a single electron of charge e, the Hamiltonian of the system

$$\hat{H} = \frac{1}{2m_e}\left[\hat{p} - \frac{e}{c}\vec{A}(\vec{r})\right]^2 \tag{2-11}$$

describes the interaction of a (spin-less) electron with an electromagnetic field. The field is given by the vector potential $\vec{A}(\vec{r})$ while $\hat{p}$ denotes the quantum mechanical momentum operator $\hat{p} = -i\hbar\nabla_r$.

Eq. 2-11 expands to:

$$\hat{H} = \frac{1}{2m_e}\left[\hat{p} - \frac{e}{c}\vec{A}(\vec{r})\right]^2 = \frac{\hat{p}^2}{2m_e} + \frac{e}{2m_e c}\left(\vec{A}\cdot\hat{p} + \hat{p}\cdot\vec{A}\right) + \frac{e^2}{2m_e c^2}\vec{A}^2 = \hat{H}_0 + \hat{H}_I \tag{2-12}$$

with the perturbation operator $\hat{H}_I$.

The *Coulomb gauge* [89] yields $\nabla_r \vec{A}(\vec{r}) = 0$ and the operator $\hat{H}_I$ in Eq. 2-11 simplifies to:

$$\hat{H}_I = \frac{e}{m_e c}\vec{A}(\vec{r})\cdot\hat{p} \tag{2-13}$$

for small perturbations, i.e. $\vec{A}^2 \approx 0$. Typically, the vector potential is described as plane wave:
$\vec{A}(\vec{r}) = A_0 \vec{e}\cdot e^{i(\vec{k}\vec{r}-\omega t)} + c$ with $\vec{e}$ being a vector of length 1 and c a complex conjugated constant.

The probability for a direct transition of an electron, being in the state $\vec{k}$ and originating from band j, to the state $\vec{k}'$ in the band j' can be expressed as (*Fermi's Golden Rule*):

$$W(j,j',\vec{k},\omega,t) = \frac{e^2 A_0^2}{m_e^2 c^2}\left|M_{jj'}(\vec{k})\right|^2 2\pi\hbar t \partial(E_{j'}(\vec{k}) - E_j(\vec{k}) - \hbar\omega). \tag{2-14}$$

$M_{jj'}(\vec{k})$ denotes the matrix element for the corresponding transition. The total number of transitions per unit of volume and time can be calculated by integration over the first Brillouin-zone:

$$W(\omega) = \sum_{jj'} \frac{1}{t} \frac{2}{(2\pi)^3} \int W(j,j',\vec{k},\omega,t) d\tau_{\vec{k}} . \qquad (2\text{-}15)$$

Eq. 2-15 has to be calculated for all occupied and unoccupied bands, j, j', and expresses the probability by which a photon of energy hv is absorbed by the solid.

The quantum mechanical term of Eq. 2-15 can now be compared with the loss of energy of a plane electromagnetic wave incident on a solid:

$$\varepsilon_2 = \frac{4\pi c^2 \hbar}{\omega^2 A_0^2} W_a(\omega). \qquad (2\text{-}16)$$

Using Eqs. 2-14, 2-15 and 2-16, energy loss and the process of photon absorption can be related to each other:

$$\varepsilon_2 = \frac{8\pi^2 c^2 \hbar^2}{m_e^2 \omega^2} \sum_{jj'} \frac{2}{(2\pi)^3} \int |M_{jj'}(\vec{k})|^2 \partial(E_{j'}(\vec{k}) - E_j(\vec{k}) - \hbar\omega) d\tau_{\vec{k}} . \qquad (2\text{-}17)$$

Applying a Jacobi-transformation, Eq. 2-17 can be expressed as integral over constant-energy surfaces:

$$\varepsilon_2 = \frac{8\pi^2 c^2 \hbar^2}{m_e^2 \omega^2} \sum_{jj'} \frac{2}{(2\pi)^3} \int_{E_{j'} - E_j = \hbar\omega} |M_{jj'}(\vec{k})|^2 \frac{df}{|\nabla_{\vec{k}}(E_{j'}(\vec{k}) - E_j(\vec{k}))|} . \qquad (2\text{-}18)$$

If the matrix elements $M_{jj'}(\vec{k})$ exhibit only small variations, i.e. $M_{jj'}(\vec{k}) \approx \text{constant}$, the integral in Eq. 2-18 can be evaluated by:

$$\varepsilon_2 = \frac{8\pi^2 c^2 \hbar^2}{m_e^2 \omega^2} \sum_{ij} |M_{jj'}(\vec{k})|^2 Z_{ij}(\omega) \qquad (2\text{-}19)$$

where $Z_{ij}(\omega)$ represents the optical joint density of states.

With $Z_{ij}(\omega) = \int_{E_j - E_i = \hbar\omega} \frac{df}{|\nabla_{\vec{k}}(E_j(\vec{k}) - E_i(\vec{k}))|}$ only terms contribute to Eq. 2-19 with vanishing denominator $\nabla_{\vec{k}}(E_j(\vec{k}) - E_i(\vec{k}))$. These singularities are called *van-Hove singularities* and correspond to the critical point energies within the energy spectrum of a semiconductor.

Finally, the physical meaning of the components of the dielectric function can be described as follows: ε_2 is directly related to the band structure of the solid and reflects the absorption properties in dependence of the irradiation energy; for low absorption ($\varepsilon_2 \sim 0$), i.e. for photon energies below the bandgap, the real part of ε is connected with the index of refraction by $\varepsilon = n^2$.

2.1.2 Brewster-angle analysis of multi-layer systems

The equations of Fresnel, expressing the reflectance behavior of a solid, can be derived from Maxwell's equations for electromagnetic waves obeying the boundary conditions at the interface where reflection occurs [89]. If a plane electromagnetic wave is considered, propagating in the (x-y)-plane along the vector $\vec{k}_i$, with velocity v_i, the electric field can be divided in components parallel (p-polarized) and perpendicular (s-polarized) to the plane of incidence (the (x-z)-plane). The respective amplitudes are denoted as $E_{i\parallel}$ and $E_{i\perp}$, respectively. If reflection occurs at a smooth interface of an isotropic medium, characterized by the dielectric function ε_S, the incident wave is partially reflected and partially transmitted. The components of the reflected wave, in turn, can be described with respect to the plane of incidence. The corresponding amplitudes are $E_{r\parallel}$ and $E_{r\perp}$. The vector $\vec{k}_r$ denotes the direction of propagation of the reflected, $\vec{k}_t$ the corresponding direction of the transmitted wave. With φ_i as angle of incidence, φ_r as angle of the reflected wave and φ_t as angle of the transmitted wave, light propagation at the air/solid interface can be illustrated as in Fig. 2-1.

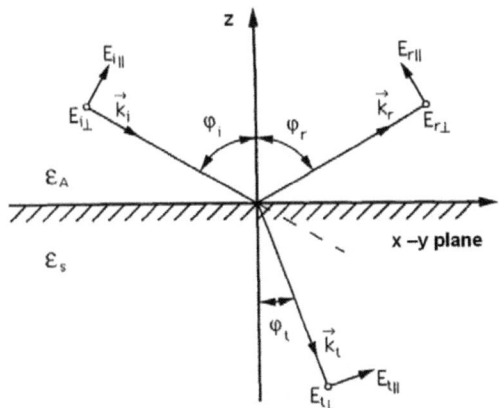

Fig. 2-1: Schematic drawing of reflection and transmission of a planar wave incident at the interface between medium A (the ambient) and S (the sample). The indices i, r and t refer to the incident, reflected and transmitted wave, respectively.

Horizontal and vertical vector components of the respective incident, reflected and transmitted wave are restrained by continuous boundary conditions. From these conditions, the fraction of the amplitude of the reflected wave with respect to the corresponding amplitude of the incident wave results in the (complex) reflection coefficients, r_p and r_s, described by:

$$E_{r\|} = r_p E_{i\|} \qquad \text{and} \qquad (2\text{-}20)$$

$$E_{r\perp} = r_s E_{i\perp}. \qquad (2\text{-}21)$$

For a semi-infinite sample, i.e., a sample with one interface only, these coefficients can be expressed for any dielectric matter and arbitrary angle of incidence, φ, as:

$$r_p = \frac{E_{r\|}}{E_{i\|}} = \frac{\varepsilon \cos\varphi - \sqrt{\varepsilon_A}\sqrt{\varepsilon - \varepsilon_A \sin^2\varphi}}{\varepsilon \cos\varphi + \sqrt{\varepsilon_A}\sqrt{\varepsilon - \varepsilon_A \sin^2\varphi}} \qquad (2\text{-}22)$$

and

$$r_s = \frac{E_{r\perp}}{E_{i\perp}} = \frac{\cos\varphi - \sqrt{\varepsilon_A}\sqrt{\varepsilon - \varepsilon_A \sin^2\varphi}}{\cos\varphi + \sqrt{\varepsilon_A}\sqrt{\varepsilon - \varepsilon_A \sin^2\varphi}}. \qquad (2\text{-}23)$$

Furthermore, it can be shown that the incident, reflected and transmitted light beams are all located within a planar area which defines the plane of incidence. The reflectance of matter can finally be calculated by the product of the complex conjugated reflection coefficients:

$$R_p(\varphi) = r_p(\varphi) \cdot r_p^*(\varphi) = |r_p(\varphi)|^2 \qquad (2\text{-}24)$$

and

$$R_s(\varphi) = r_s(\varphi) \cdot r_s^*(\varphi) = |r_s(\varphi)|^2. \qquad (2\text{-}25)$$

For a surface exposed to air it is common practice to set $\varepsilon_A = 1$ corresponding to vacuum conditions where the relative dielectric constant equals to 1. Eq. 2-22 simplifies then to:

$$r_p(\varphi) = \frac{\varepsilon \cos\varphi - \sqrt{\varepsilon - \sin^2\varphi}}{\varepsilon \cos\varphi + \sqrt{\varepsilon - \sin^2\varphi}} \qquad (2\text{-}26)$$

while Eq. 2-23 changes correspondingly.

In order to determine the optical properties of a solid, i.e., the two components of the complex dielectric function ε, two independent quantities have to be measured. Brewster-angle analysis (BAA) is based on the simultaneous measurement of the Brewster angle φ_B and the reflectance R_p at this angle [92-94].

BAA determines the minimal reflectance $R_p(\varphi_B)$ at the Brewster angle φ_B for p-polarized light, i.e., φ_B is defined by the condition $\dfrac{dR_p(\varphi)}{d\varphi}=0$, where R_p is given by the square of the absolute value ($R_p = r_p \cdot r_p^*$, see Eq. 2-22) of the complex reflection coefficient for the material with dielectric constant $\varepsilon = \varepsilon_1 + i\varepsilon_2$. The behavior of the reflectance of silicon for a photon energy $h\nu = 2.48$ eV, corresponding to 500 nm wavelength, is exemplified in Fig. 2-2. The inset represents a magnification around the Brewster-angle of $\varphi_B \sim 76.9°$.

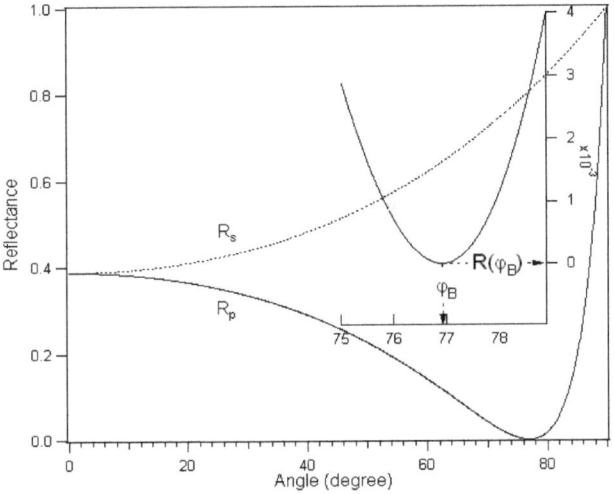

Fig. 2-2: Reflectance behavior of silicon for varied angles of incidence for perpendicular and parallel polarized light, R_s and R_p, respectively. The calculation is based on published values for the dielectric function [95] for a photon energy of $h\nu \sim 2.48$ eV corresponding to a wavelength of $\lambda = 500$ nm. The inset shows the behavior of R_p around the Brewster angle where the reflectance almost vanishes.

The relationship of the Brewster angle to the dielectric function of a solid has been calculated [96] as:

$$\varphi_B = \arcsin\left[\sqrt{\dfrac{-|\varepsilon|^2}{3(|\varepsilon|^2+\varepsilon_1)}}\left[|\varepsilon|^2 - 3 + \cos\left(\dfrac{\chi}{3} + \dfrac{4\pi}{3}\right)\sqrt{|\varepsilon|^4 + 6|\varepsilon|^2 + 12\varepsilon_1 + 9}\right]\right], \qquad (2\text{-}27)$$

where $\cos\chi = \dfrac{|\varepsilon|^4 \left(|\varepsilon|^8 + 9|\varepsilon|^6 + 27|\varepsilon|^4 + 18|\varepsilon|^2 \varepsilon_1 - 27|\varepsilon|^2 + 54|\varepsilon|^2 \varepsilon_1 + 54\varepsilon_1^2\right)}{\sqrt{(|\varepsilon|^4 + 6|\varepsilon|^2 + 12\varepsilon_1 + 9)^3}}$.

So far, a semi-infinite sample, corresponding to a two-phase system (ambient-sample), was considered. In practice, the interface of a sample towards the ambient is characterized by dielectric properties that deviate from those in the bulk. Surface relaxation and reconstruction result in modified optical properties at the surface. Even more influential are atomic roughness and possible surface contamination arising from the ambient. All these effects constitute a multi-layer system that requires modification of the equations given above.

In a first approximation, surface effects can be comprised by the assumption of an additional thin layer that causes multiple reflections at the respective interfaces as illustrated in Fig. 2-3:

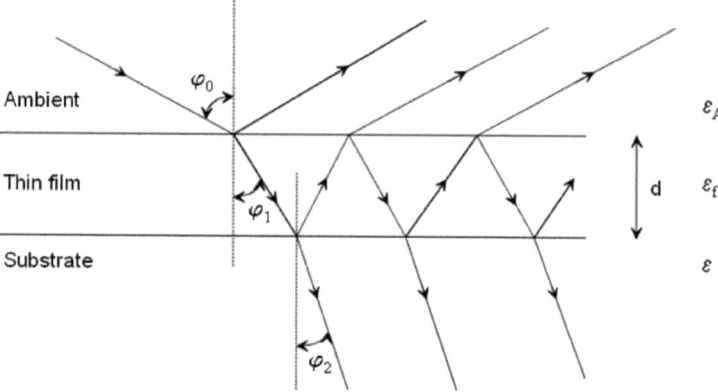

Fig. 2-3: Multiple reflections of an incident light beam at the interfaces of a substrate covered by a thin film. This film-substrate system represents a three-phase model consisting of the ambient, a thin film and the substrate and, correspondingly, two interfaces.

For these more complex systems, Fresnel's equations for multilayer structures have to be used [97] which are given in the following in their most general form for an arbitrary number of films:

$$\begin{pmatrix} M_{11} & M_{12} \\ M_{21} & M_{22} \end{pmatrix} =$$

$$= \begin{pmatrix} 1 & r_{a1} \\ r_{A1} & 1 \end{pmatrix} \begin{pmatrix} 1 & r_{12} \\ r_{12}e^{-2i\Phi_1} & e^{-2i\Phi_1} \end{pmatrix} \begin{pmatrix} 1 & r_{23} \\ r_{23}e^{-2i\Phi_2} & e^{-2i\Phi_2} \end{pmatrix} \times \ldots \times \begin{pmatrix} 1 & r_{(n-1)n} \\ r_{(n-1)n}e^{-2i\Phi_{n-1}} & e^{-2i\Phi_{n-1}} \end{pmatrix}. \quad (2\text{-}28)$$

Here, the resulting complex reflection coefficient is given by:

$$r_p = \frac{M_{21}}{M_{11}}, \quad (2\text{-}29)$$

where $r_{i(i+1)}$ describes the respective reflection coefficient at the boundary between the ith and the $(i+1)^{th}$ layer:

$$r_{i(i+1)_p} = \frac{\varepsilon_{(i+1)}\sqrt{\varepsilon_i - \varepsilon_A \sin^2 \varphi} - \varepsilon_i\sqrt{\varepsilon_{(i+1)} - \varepsilon_A \sin^2 \varphi}}{\varepsilon_{(i+1)}\sqrt{\varepsilon_i - \varepsilon_A \sin^2 \varphi} + \varepsilon_i\sqrt{\varepsilon_{(i+1)} - \varepsilon_A \sin^2 \varphi}}. \quad (2\text{-}30)$$

The exponent $\Phi_i = \frac{2\pi d_i}{\lambda}\sqrt{\varepsilon_i - \varepsilon_A \sin^2 \varphi}$ accounts for the phase shift at the respective boundary and depends upon the optical constants ε_i, ε_A and the layer thickness d_i of the ith layer.

Real semiconductor surfaces or systems of layers always show some corrugation, at least on the atomic level. This deviation from a smooth topography affects both the surface, exposed to the impinging light beam, and all interfaces. In order to take these perturbations into account, a number of models were derived for post-experimental analysis of the optical data. In many cases the averaging of the dielectric properties of two or more phases is sufficient to obtain an *effective dielectric function* of the specimen. This approach, called *effective medium approximation* (EMA), is valid as long as the feature size of the pore geometries is much smaller than the wavelength of the probing light and so-called retardation effects can be neglected. Several models are known from which effective dielectric functions can be derived. Since proper 'mixing-rules' (for the phases) are not easy to define, the models differ by complexity and applicability. In the Bergman theory [98-100],

$$\varepsilon_{eff} = \varepsilon_M \left(1 - v \int_0^1 \frac{g(n,v)}{\frac{\varepsilon_M}{\varepsilon_M - \varepsilon} - n} dn \right), \quad (2\text{-}31)$$

the spectral density function g(n) models geometrical resonances of two materials with dielectric functions ε_M and ε. The function g(n,v), obeying the condition $\int_0^1 g(n,v)dn = 1$, is independent of the material properties. Eq. 2-31 can therefore be regarded as the most general effective medium approach.

The Maxwell Garnett model [101] is given by:

$$\frac{\varepsilon_{eff} - \varepsilon_M}{\varepsilon_{eff} + 2\varepsilon_M} = (1-v)\frac{\varepsilon - \varepsilon_M}{\varepsilon + 2\varepsilon_M}. \quad (2\text{-}32)$$

This model considers sharp geometrical resonances and is suited to simulate either large porosities or rough surfaces with spherical objects of large mutual distance. This model is applied in section 3.1 where a successively etched Si(111) surface is almost smooth with only small remnants of surface corrugation.

The Bruggeman formula [102] is the most frequently used:

$$v \frac{\varepsilon_M - \varepsilon_{eff}}{\varepsilon_M + 2\varepsilon_{eff}} + (1-v) \frac{\varepsilon - \varepsilon_{eff}}{\varepsilon + 2\varepsilon_{eff}} = 0. \qquad (2\text{-}33).$$

Eq. (2-33) represents the case of continuous distribution of resonances and is applied in sections 3.1 and 3.2 for larger, irregular surface roughness.

In the formulas shown above ε_{eff} refers to the resulting effective dielectric constant of the mixed phase comprising a v-fraction of material with dielectric constant ε (the host material) and (1-v)-fraction of material characterized by the dielectric function ε_M (the inclusions).

Fig. 2-3 shows the schematic of the experimental BAA setup. Optical components such as Tungsten-Halogen light source (LS), monochromator (M), chopper (C) and a diaphragm are arranged in order to direct the light beam to a beam splitter (BS). As monochromator, a ½ m - Czerny-Turner-monochromator (Digikröm) is used whose large focal distance allows high resolution measurements but also reduces the intensity of the probing light beam. The beam splitter is employed for selection of about 10% of the total light intensity in order to realize a reference light beam. This reference beam is used to minimize the influence of fluctuations of the light source on the data and is detected by the cooled silicon photodetector D_1 (Hamamatsu). The probing light is directed through a Glan-Thompson polarizer with an extinction ratio of 10^{-6} and is detected, after reflection, by the photodetector D_2. Both photodetectors are connected each to a Keithley current-voltage amplifier (A_1, A_2). The sample (S) is attached to a sample holder fixed to a goniometer table with angle-step accuracy of 0.004°. During the detection cycle, reflectance values of the parallel polarized light are measured within a predefined interval of $\pm 2°$ around the Brewster angle of the sample with a step-width of 0.02°.

Lock-in technique (EG&G) is used to reduce noise effects. The chopper frequency of 277 Hz serves herby as reference input for the Lock-in devices (L_1, L_2). The measured data are then divided by the instrumental work function obtained by an initial measurement of the light signal directly incident on the photodetector D_2. The parabola-like behavior of the reflectance around the Brewster angle (see Fig. 2-2) is analyzed by a polynomial fitting routine.

Polynomials of second, third or fourth degree can be chosen by the experimenter. This choice was implemented into the data recording and evaluation software (LabVIEW, National Instruments) in order to realize either smooth spectral curves (applying polynomials of second degree) or Brewster angle and reflectance values with increased accuracy (applying polynomials of fourth degree).

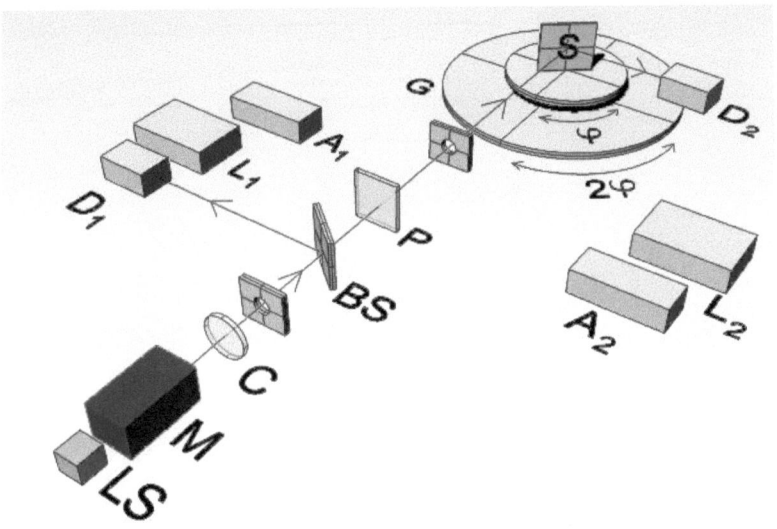

Fig. 2-3: Experimental setup for Brewster-angle analysis (BAA) measurements with components: LS (light source), M (monochromator), C (chopper), BS (beam splitter), $L_{1,2}$ (Lock-In amplifier), $A_{1,2}$ (current-voltage amplifier), $D_{1,2}$ (photo-detectors), P (polarizer), G (two-stage goniometric table). S denotes the sample fixed to the sample holder.

The BAA data evaluation requires additional consideration of technical imperfections which arise, e.g., from light beam divergence and are affecting the accuracy of the polarization state and the accuracy of the angle of incidence. The corresponding mathematical treatment of these effects are presented in appendix A.1 together with the routines which were applied for numerical computation of the quantities $R_p(\varphi_B)$ and φ_B in multi-layer analysis.

2.2 *In situ* Brewster-angle reflectometry of (electro-)chemical conditioned silicon surfaces

2.2.1 Experimental arrangement

The instrumentation of *in situ* Brewster-angle reflectometry (BAR) follows the BAA setup shown in Fig. 2-3. Additionally, the goniometer table is equipped with a BF_2 glass beaker, resistant to electrolyte corrosion at least for the time of one experiment, an electrode system and a potentiostat (see Fig. 2-4).

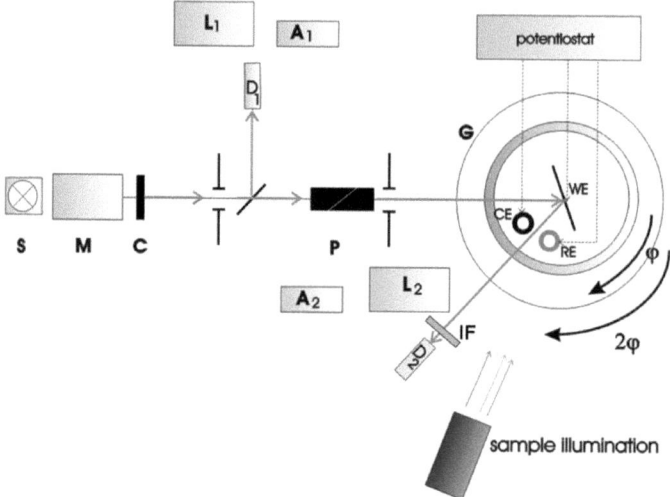

Fig. 2-4: Experimental setup for *in situ* Brewster-angle reflectometry (BAR) measurements with abbreviations: S (light source), M (monochromator), C (chopper), $L_{1,2}$ (Lock-In amplifier), $A_{1,2}$ (current-voltage amplifier), $D_{1,2}$ (photo-detectors), P (polarizer), G (two-stage goniometric table), IF interference filter (500 nm), WE working electrode (sample), CE counter electrode, RE reference electrode.

As current-voltage detection system, a three-electrode configuration was used comprising a Pt-counter electrode, a Ag/AgCl reference electrode while the sample, mounted with InGa amalgam to a Mo plate and enclosed within a VITON O-ring, served as working electrode. For photoelectrochemical conditioning, an external W-I lamp was used with adjustable light intensities between 0.4 $\mu W/cm^2$ to 50 mW/cm^2. In order to prevent this light from reaching the detector, an interference filter was positioned in front of the detector with small band-width around the photon energy of the probing light (500 nm). In contrast to conventional reflectometry, the angle of incidence was adjusted to the Brewster angle of the silicon substrate or, for ultra-thin films, to the effective Brewster angle of the film-substrate system.

The wavelength of 500 nm, corresponding to 2.48 eV, was chosen to optimize the relative increase of the reflectance signal in the presence of surface roughness. During electrochemical experiments, this angle was permanently kept at its initial value to provide fast data acquisition at a rate of about 2 data points per second. The intensity of the probing light (about 4μ Wcm^{-2}) was too low to influence the reflectance data by unintended photoelectrochemical reactions.

2.2.2 The linear approximation of the reflectance

Real-time monitoring of changing surface conditions benefits from straightforward interpretation of the measured data. Although multi-layer analysis, as explained in the preceding section, requires consideration of the intricate interdependence of reflection contributions from several interfaces, a simplified evaluation of the data can be applied which allows real-time manipulation of experimental parameters during the experiment (see section 3.2). In Fig. 2-5, the reflectance response of initially H-terminated silicon, enclosed into a glass beaker filled with water, is shown for different variations of the surface condition: A) for increasing thickness of an oxidized layer, B) for increasing thickness of a rough layer and C) for increasing porosity of a 5 nm porous layer:

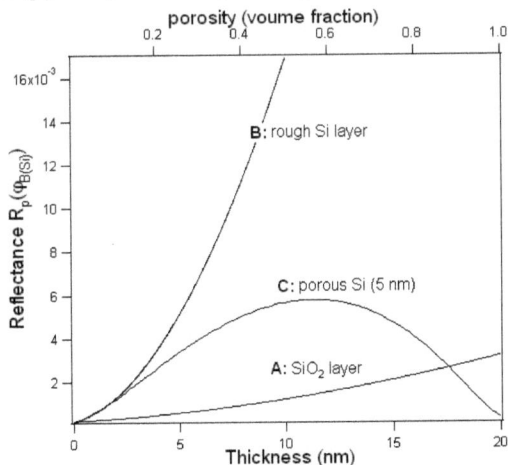

Fig. 2-5: Calculated response of $R_p(\varphi_B)$ to the presence of increasing surface roughness, of a 5 nm thick porous Si layer with increasing porosity and of a SiO$_2$ oxide layer with increasing thickness.

The data were calculated according to Bruggeman effective medium approximation as introduced before. Additionally, instrumental imperfections were included in the computations. Therefore, the shown values represent data as they would be measured by the BAR system. Curve A shows the effect of an oxide layer of increasing thickness (the index of refraction of the surrounding liquid is 1.33). In the range between 4 nm and 20 nm, the resulting reflectance curve behaves, in first approximation, linearly and the BAR reflectance signal would increase/decrease proportional to the growth/etching of the oxide layer. Curve B illustrates the case of increasing surface roughness, modeled by an appropriate effective dielectric function with an ambient/bulk ratio v = 0.5. In contrast to these curves, curve C is calculated for a porous layer of constant thickness and varying porosity, buried beneath the electrolyte-surface interface. In this case, a certain void to bulk fraction exists where the reflectance curve is inflected. In order to distinguish interfering surface effects, real-time interpretation therefore requires some knowledge about possible processes which are determinant at a given potential. Then experimental parameters as potential, light intensity or electrolyte composition can be manipulated during the conditioning process in order to achieve the desired chemical and/or topographical surface state.

2.3 Photoelectron spectroscopy using synchrotron radiation

Synchrotron radiation bridges the gap between ultra-violet spectroscopy and X-ray photoelectron spectroscopy available as laboratory instrument (using an anode of fixed emission energy such as Mg K_α). Since applications of synchrotron radiation are numerous [103], the following survey will focus on the main application being of interest for this work: the tunable excitation energy that allows high-resolution elemental analysis of surfaces.

2.3.2 Principles of photoelectron excitation

X-ray photoelectron spectroscopy (XPS) exploits the fact that photoemission of an electron of binding energy E_B is specific to a chemical element [104]. A simplified quantum mechanical approach for a single atom [105] shows that the binding energy is a function of the quantum number n and the charge of the nucleus Z:

$$E_B = -(13.6eV)\frac{Z^2}{n^2}. \qquad (2\text{-}34)$$

Generally, the emission process of an electron from condensed matter can be conceived as three-step sequence [106]:

1. The excitation by absorption of a photon
2. The transport towards the surface
3. The escape into the vacuum.

In Fig. 2-6a the excitation of a core-level electron, bound to the K-shell, is shown that bears enough energy after excitation to escape into the vacuum.

In the energy relation for the resulting kinetic energy of the escaping photoelectron

$$E_{kin} = hv - E_B - \Phi \qquad (2\text{-}35)$$

only the first and third step are considered: the photon of energy hv transfers its energy to a core-level electron; some energy (the binding energy E_B) is required to leave the atom; some other energy (the work function Φ) is required to leave the solid. Step 2 finally may lead to inelastic scattering which appears then in the corresponding XPS spectrum as unstructured background. The energy term according to Eq. 2-35 is called *Koopman's energy*. The fact that measured photoelectron energies considerably deviate from this relation will be discussed further below.

The hole that is left behind after photoelectron emission can be occupied by another electron from a higher electron shell as illustrated by Fig. 2-6:

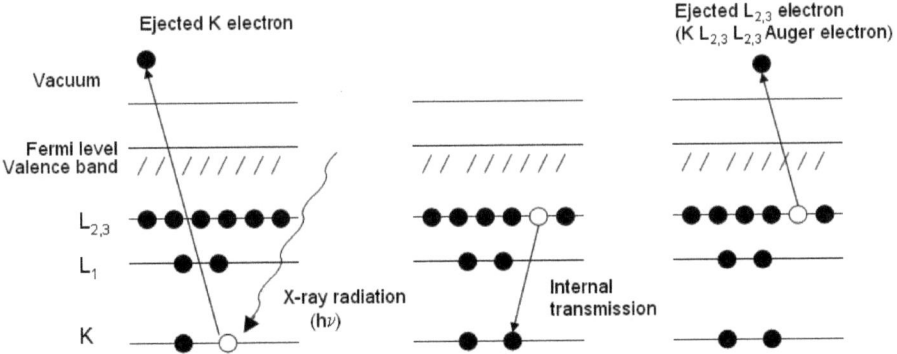

Fig. 2-6: Excitation of a core-level electron (K-shell) after absorption of a photon of energy hv: (a) Emission of the electron into the vacuum. (b) Occupation of the hole by another electron and transfer of energy to another L-shell electron. (c) Emission of the L-shell electron (Auger electron), according to [107].

By this process, the electron will gain some energy which can, in part, subsequently be transferred to a third electron which then will be emitted as so-called Auger electron. The term KLL-emission refers to the multiple-step process involving three electrons from two different atomic shells. Auger electrons originate usually from the L-shells of an atom. Consequently, their binding energies are lower but independent from the excitation energy hv. This makes Auger electron spectroscopy a useful technique for chemical analysis.

Necessary corrections of Eq. (2-35) are based on the following considerations: after photoelectron emission from the n-electron system, fast reconfiguration of the (n-1)-electron system leads to screening of the remaining hole. This relaxation process leads to an equilibrium state of lower energy. The energy gain, $E_{n(n-1)}$, is transferred to the emitted electron which is then measured at higher kinetic energies (lower binding energies). In the *sudden approximation* model, only the relaxation term $E_{n(n-1)}$ is added to the right-hand side of Eq. 2-33. This approximation refers to the adiabatic limit that states that photoionization and photoemission are slow (compared to relaxation effects) such that the system is always in equilibrium.

However, the actual photoemission process is very fast (< 1 fs) and the perturbation of the (n-1)-system can lead to excited final states (*shake-up satellites*) or continuum states (*shake-off satellites*) in the measured spectrum. Most important for the chemical analysis are *extrinsic* contributions to the relaxation term which result from the total atomic configuration in the vicinity of the emitting atom. Neighboring atoms of different chemical origin alter the resulting kinetic energy of the emitted electron by a term summarized as the *chemical shift*, E_{chem}. Thus, Eq. (2-35) has to be completed by, at least, two additional terms:

$$E_{kin} = hv - E_B - E_{chem} - E_{\Delta(n-1)} - \Phi . \qquad (2-36)$$

In solids, core-level energies are detected by discrete values. Electrons, related to the valence band, are weakly bound and have energies distributed over a certain range. In this case, the excitation of an electron, being represented as quantum mechanical wave function ψ_i, in its initial state and ψ_f in its final state, can be expressed using the transition probability:

$$\mu_{fi} = \int \psi_f H_{PE} \psi_i dr = <\psi_i | H_{PE} | \psi_f> . \qquad (2-37)$$

Here, the perturbation Hamiltonian H_{PE} for photoemission can be derived according to section 2.1.1 (Eq.2-12). Using the linear (dipole) approximation, $H_{PE} = \frac{e}{m_e c} \vec{A} \cdot \vec{p}$, the resulting photocurrent in $\vec{r}$-direction can by expressed according to *Fermi's Golden Rule*:

$$I(\vec{r},E_f,h\nu) = \frac{2\pi e}{\hbar}\left(\frac{e}{m_e c}\right)^2 \sum_i |\mu_{fi}|^2 (E_f - E_i - h\nu). \quad (2\text{-}38)$$

In terms of the *sudden approximation* model, the photoelectron in the final state, ψ_f, is decoupled from the remaining solid, i.e. any *extrinsic* interaction is excluded after photoexcitation. This approach permits simplified evaluation of the expression 2-38.

The number of excited electrons, on the other hand, is given by:

$$N(\vec{r},E_f,h\nu) \sim \int D_i(E_i) D_f(E_f)(E_i + h\nu) |\mu_{fi}|^2 \, dE_i . \quad (2\text{-}39)$$

Since Eq. 2-39 uses the joint density of states, the densities of initial and final states likewise contribute to the resulting energy dispersive curve (EDC). In order to disentangle the respective contributions, variation of the photoelectron excitation and detection procedure can be applied: in the constant final state mode (CFS), the kinetic energy is recorded at a fixed value while the excitation energy is varied within a certain energy range; in the constant initial state mode (CIS), both the detected kinetic energy and the excitation energy are synchronously changed. Hence, the resulting spectra are mostly determined by the density of initial and, respectively, final states.

In practice, the measured kinetic energies are influenced by instrumental restrictions. The Mg K$_a$ X-ray source is of finite width of approximately 0.7 eV. Therefore, broadening of the detected core-level peaks appears in the spectra which hinders, for instance, detection of the spin-orbit splitting of the Si 2p core level if the X-ray source is not monochromatized by application of suitable diffraction gratings. Higher resolution is given by sources used in Ultraviolet Photoelectron Spectroscopy (UPS) such as gas discharge lamps using Helium ion emission (He I ≙ 21.2 eV, He II ≙ 40.8 eV). The He II line, e.g., has a line width of about 80 meV. The spectrometer, on the other hand, contributes to the emission process by the spectrometer work function, i.e., after leaving the specimen, electrons have to overcome the energetic barrier given by the solid state components of the XPS instrument. Calibration of the spectrometer, prior to sample investigation, is therefore required in order to analyze this additional contribution to the work function. It is common practice to employ well defined metal surfaces such as Au to correlate the detected core-levels to published values (the Au 4f lines at about 84 eV and 88 eV). However, nonlinearities in the spectrometer work function have to be considered and the calibration procedure has to be measured at different points of the full energy range. This can be carried out by measuring Au, Ag and Cu at the energies of about 84 eV, 368.3 eV and 932.7 eV and subsequent nonlinear optimization of the spectrometer work function.

In Fig. 2-7, the detection of photoelectrons is schematically shown for core-level as well as valence band electrons. The assumed work functions of the specimen and the instrumental energy barrier are indicated. As a result of the combined contributions of the densities of states, related to initial and final wave functions, the shape of the measured valence band differs from that of the solid in the resulting energy dispersive curve. Inelastic scattering of electrons introduces the background signal with a steep cut-off at the lowest boundary of the kinetic energy interval ($E_{kin} = 0$).

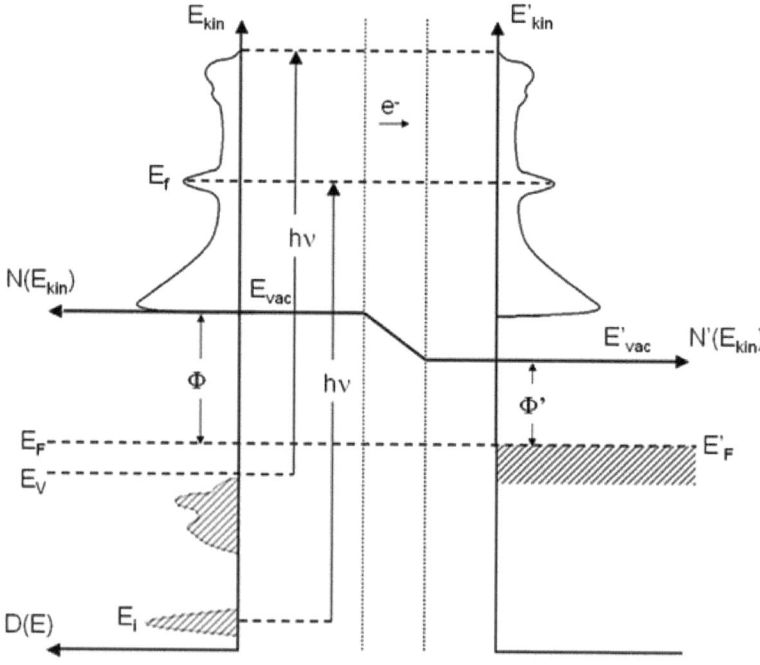

Fig. 2-7: Photoelectron emission of valence band electrons and from core-levels (E_i) after excitation by light with photon energy hv. Fermi levels of the semiconductor (E_F) and the spectrometer (E'_F) are aligned. Shaded areas describe the electron densities, D(E), of the semiconductor while N(E) denotes the number of detected photoelectrons.

The cut-off of scattered electrons with kinetic energy $E_{kin}=0$ (secondary electrons) is an important indicator for the presence of dipole layers at the surface. Such dipoles are arising, for instance, from electrons outside a metal or semiconductor. The corresponding wave functions exhibit a decay that extends over the boundary phase, thereby introducing negative charges in front of the solid. During measurement, these dipoles are changing the work function and are detected by a shift of the cut-off energy, $\Delta\chi$, which lowers or raises the electron affinity χ [104]. Other sources of surface dipoles are adsorbates which are characterized by partial charging. This is illustrated in Fig. 2-8b and compared to the UV-spectrum in absence of a dipole layer (Fig. 2-8a).

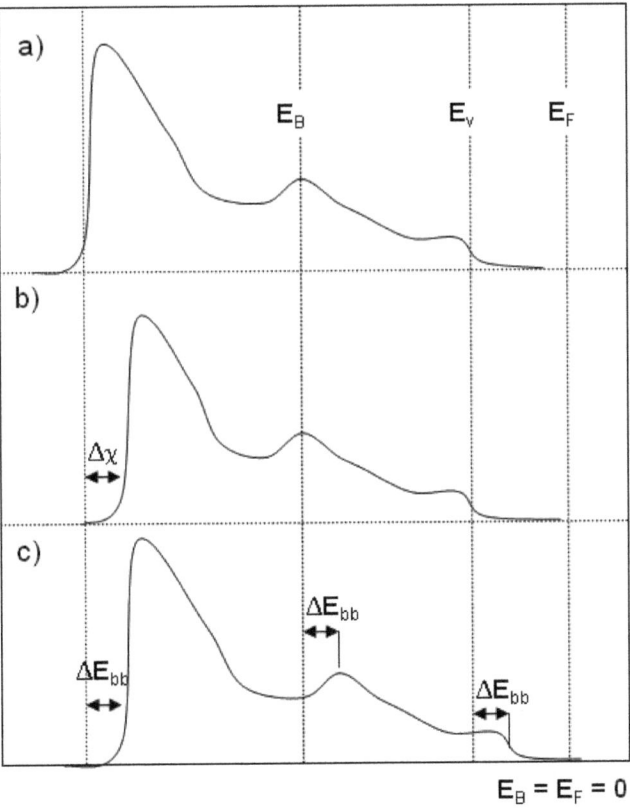

Fig. 2-8: Influence of band bending and surface dipole layers on the photoelectron emission spectrum after excitation with a UV-light source. (a) Typical UV spectrum with a core level energy at E_B, valence band position E_V, and Fermi energy E_F. (b) Shift $\Delta\chi$ of the cut-off of secondary electrons due to a surface dipole layer. (c) Shift of the whole spectrum due to band bending, ΔE_{bb}.

A second effect, to be distinguished from dipole layers, is band banding in semiconductors. By reaching an equilibrium state between the metal back-contact and electronically active surface states, the position of the Fermi-level at the surface differs from the bulk position. Band bending introduces a shift of the whole energy dispersive curve, i.e. the spectrum width is preserved. Both effects are illustrated by Fig. 2-8 where a spectrum is shown without surface dipoles or band bending (Fig. 2-8a), in presence of a dipole layer (Fig. 2-8b) and influenced by band bending (Fig. 2-8c).

2.3.2 The application of synchrotron radiation

In synchrotrons large bending magnets are employed to accelerate electrons in circular paths. During acceleration, the electric field associated with the charged particles exhibits rearrangement of the distribution of its lines of force. These field perturbations result in electromagnetic radiation. The radiated power can be derived from *Larmor's formula* [108] for a non-relativistic accelerated charge (corresponding to Hertz' dipole radiation [109]):

$$P = \frac{2}{3}\frac{e^2}{4\pi\varepsilon_0 c^3}a^2 \ . \tag{2-40}$$

Since electrons in synchrotrons are highly relativistic, the acceleration has to be substituted by:

$$a = \frac{1}{m}\frac{dp}{d\tau} = \frac{1}{m}\gamma\frac{d(\gamma m v)}{dt} = \gamma^2\frac{dv}{dt} = \gamma^2\frac{v^2}{r} \ , \tag{2-41}$$

where $\gamma = \frac{1}{\sqrt{1-\frac{v^2}{c^2}}}$ (Lorentz factor) and $\tau = \frac{t}{\gamma}$.

The radiated power can then be expressed by [110]:

$$P = \frac{2}{3}\frac{e^2}{4\pi\varepsilon_0 c^2}\left[\gamma^2\frac{v^2}{r}\right]^2 = \frac{2}{3}\frac{e^2\gamma^4 v^4}{4\pi\varepsilon_0 c^3 r^2} \ . \tag{2-42}$$

In synchrotron facilities, large storage rings are used to maintain the circular movement of the electrons which results in persistent emission of synchrotron light. Bending magnets are used to reinforce the circular orbit while other dipole magnets are employed either to invoke sharp energy peak lines or broad energy spectra until a critical energy: in *undulators* electrons are forced on a sinusoidal path by weak magnetic fields. High brilliant synchrotron light with a small energy distribution is achieved by this process. *Wiggler* magnets, on the other hand, add higher orders to the electromagnetic spectrum providing thus an energy range with low spectral intensity variation. Both devices consist of a series of periodically aligned dipole magnets with alternating magnetic field as illustrated by Fig. 2-9:

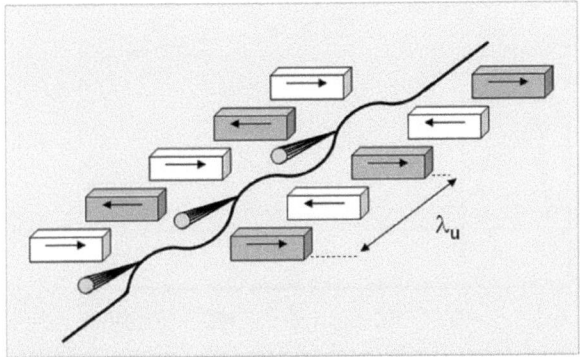

Fig. 2-9: Schematic arrangement of dipole magnets in an undulator/wiggler structure. The imposed wavelength λ_u determines the bandwidth of the emitted radiation.

The length of the dipole arrangement determines the characteristic wavelength of electron oscillation, λ_u, and the bandwidth of the radiation. With the dimensionless parameter

$$K = \frac{eB\lambda_u}{2\pi m_e c} \qquad (2\text{-}43)$$

the emitted radiation can be characterized in dependence of λ_u. For $K \ll 1$, interference patterns intensify the radiation in narrow spectral bands. This is the typical application of dipole magnets in an undulator. For $K \gg 1$, a broad spectrum is observed by independently superimposed field contributions. This mode of operation is applied by wiggler devices.

Synchrotron radiation is characterized by high brilliance, i.e. intensity per area and time unit and per percentage of the band width. The light is polarized in the plane of acceleration and is emitted in ultra-short pulses (ps range) separated by time intervals in the nano-second range. The radiated power density shows typically a distribution as depicted in Fig. 2-10. The usable energy range drops exponentially off behind a critical energy E_c (corresponding to wavelength λ_c). The critical energy is defined by [110]:

$$E_{crit} = \hbar\omega_{crit} = \frac{3\hbar\omega_0\gamma^3}{4\pi} \sim B\gamma^2. \qquad (2\text{-}44)$$

Here, ω_0 defines the Larmor frequency which depends on the magnetic field B.

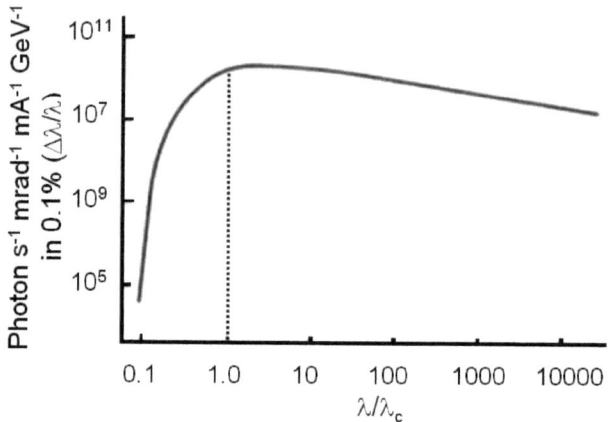

Fig. 2-10: Normalized spectral power density provided by a typical synchrotron radiation facility. The photon number is depicted in dependence of the ring current, a small angle within the emission cone and with respect to 0.1% of the band-width [111].

Storage ring architecture is exemplifying presented in Fig. 2-11 by the facilities at Bessy II, Berlin-Adlershof, Germany.

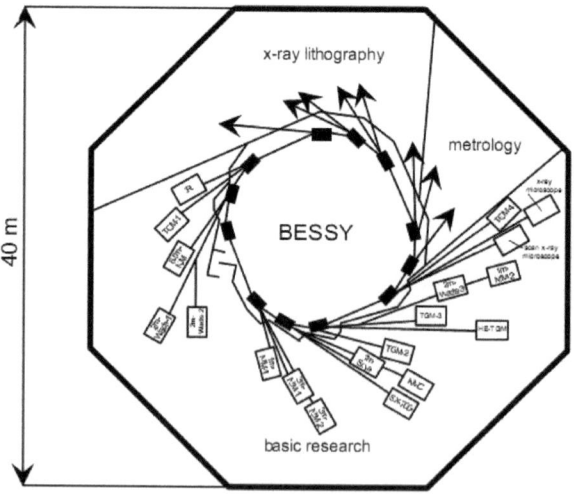

Fig. 2-11: Synchrotron facilities at Bessy II, Berlin-Adlershof, Germany. Indicated are the storage ring, beam lines with access to the synchrotron radiation and the arrangement of principle measurement areas.

Extending from the storage ring, so-called beam lines are indicated as straight lines where ray tubes, lenses and monochromators are employed to provide synchrotron light with well defined photon energy.

Application of synchrotron light offers highest resolution in time, energy and, with appropriate additional means, space. For the work presented here, the high-resolution energy range is of particular importance. While the energy resolution allows for detection of core-level substructures as the spin-orbit splitting of the Si 2p line or the distinction of silicon dioxides in substoichiometric configurations (see chapter 1.1.1), the tunable photon energy makes furthermore selective chemical analysis possible with respect to the vertical distribution within the specimen. As electrons are excited, scattering takes place during their passage towards the surface and determines the escape probability. Multiple scattering events lead only to contributions to the background spectrum while no or small scattering is visible at the corresponding core-level lines. Since scattering is related to the escape depth of the electrons which, in turn, depends on the excitation energy, depth-dependent information can be obtained by tuning of the radiation energy to an appropriate magnitude. In Fig. 2-12, measured and calculated values of the escape depth of photoelectrons in dependence of the kinetic energy are given in terms of monolayers of the specimen under investigation.

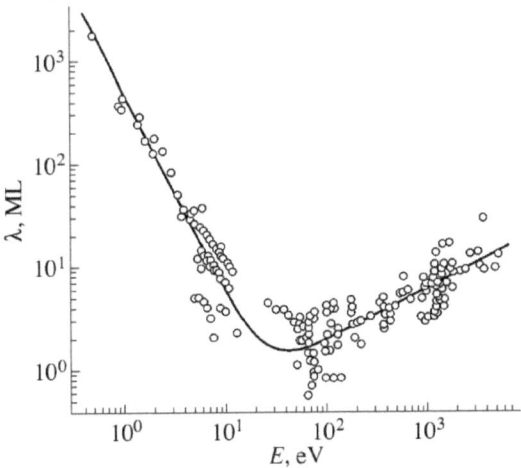

Fig. 2-12: Escape depth of photoelectrons after X-ray excitation [112]. The escape depth is indicated in monolayers in dependence on the kinetic energy after photoemission.

From Fig. 2-12 follows, that depending on the chemical element under investigation, excitation energies can be adjusted such that the kinetic energies of photoelectrons fall within the desired range of more surface or more bulk-sensitive measurements. For instance, with the Si 2p binding energy of about 99 eV, photoelectron excitation with energies of about 130 – 170 eV is related to an escape depth of about only one monolayer or about 4Å. This can be concluded from the resulting kinetic energy of the photoelectrons of about 30 – 70 eV after emission which corresponds to the minimum of the curve in Fig. 2-12.

2.4 Photoelectron emission microscopy

The first working photoemission electron microscope (PEEM) was built by E. Brüche in 1933 using ultraviolet (UV) light to image photoelectrons emitted from a metal [113]. The principal design of this PEEM apparatus is still used. G. F. Rempfer, H. H. Rotermund, G. Ertl, and G. Schönhense, beside others, developed improved electron optics in the 1980s and 1990s [114-116]. Today a spatial resolution below 10 nm can be routinely achieved using UV light and several 10 nm using X-ray excitation [117]. Rempfer et al. calculated a best resolution of 3 – 5 nm for this technique [118]. Modern correction schemes for aberration errors are about to improve the resolution down to a few nanometers, close to the physical limit of emission microscopy which is determined by the mean free path of low energy electrons.

2.4.1 Experimental arrangement

The experimental arrangement for PEEM measurements is schematically shown in Fig. 2-13. A voltage of between 15 kV and 20 kV accelerates the photoemitted electrons from the sample. The objective lens and transfer lens produce an intermediary image behind a backfocal plane aperture which is then magnified by two projector lenses. Spatial resolution and transmission (efficiency) of the electron optics can be varied using different backfocal plane apertures with sizes between 15 mm and 50 mm.

A cooled charge-coupled device (CCD), fiber-coupled to a phosphor detector monitors the electron-optical image. Chromatic aberrations (electrons with different velocity) and spherical aberrations (electrons reaching the detector at different angles) lead to blurring of the image. The hyperbolic field of a curved electron mirror can in principle correct the effect of both types of aberrations in a PEEM instrument. This is the idea behind aberration-correction techniques which have been successfully employed in light microscopes and transmission electron microscopes. Chromatic aberration dominates if X-ray excitation is used because the emitted electrons have a much larger energy spread compared to UV excitation.

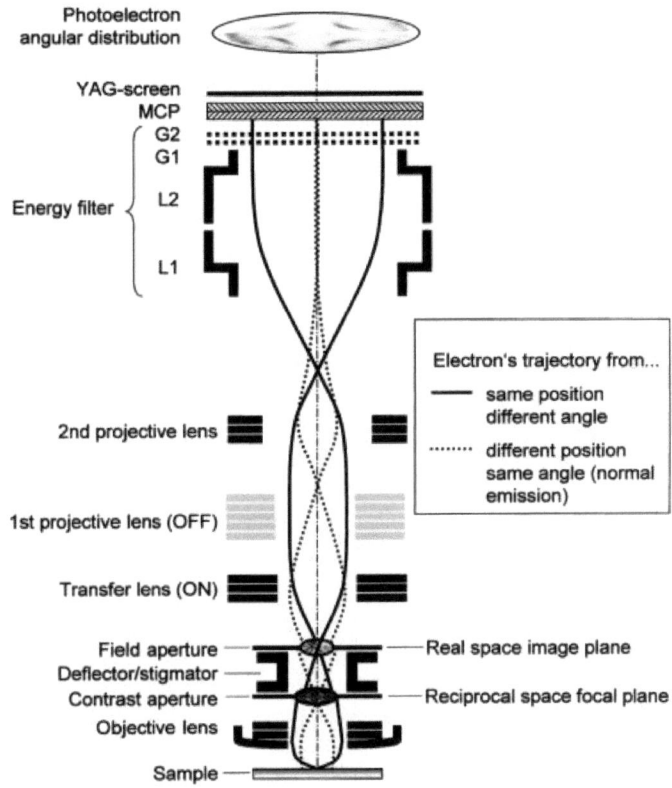

Fig. 2-13: Principle setup of a photoelectron emission microscope [119]. Components and working principles are discussed in the text.

2.4.2 Contrast in photoelectron emission microscopy

Image contrast in PEEM measurements is mainly due to chemical variation across the specimen's surface. However, the dependence of the photoionization cross section on the surface orientation of crystalline samples makes it also possible to detect, e.g., grain boundaries in polycrystalline material. Moreover, pronounced topographic variation can result in a curvature of the accelerating electric field yielding thus topographic information (at weaker spatial resolution). Surface protrusions may result in distortions of the electric field applied to the sample. These distortions deflect or accelerate electrons due to the variation of the force lines of the electric field. As a result, electrons emitted from an area of a feature which is inclined with respect to the surface normal may be blocked by the aperture and these areas appear darker in the image than flat top-areas. At sharp points, the local electric field is enhanced and these points appear brighter on the image. On the other hand, areas indented into the surface, may not be reached by the excitation light and will not contribute to the image at all. Work function differences, in turn, dominate upon UV illumination. Although the samples investigated by PEEM in this work exhibit deep etch grooves, only elemental contrast will be discussed which is comprehensible by the considerations of section 2.3.

2.5 Atomic force microscopy

Scanning Probe Microscopy (SPM) comprises a large family of microscopy techniques where a sharp probe is scanned across a surface and the probe-sample interaction is monitored. The two primary methods of this family are Scanning Tunneling Microscopy (STM, Binnig and Rohrer, 1982/1983) [120, 121] and Atomic Force Microscopy (AFM, Binnig, Quate and Gerber, 1986) [122]. The advantage of AFM over STM is that the sample does not necessarily have to be conductive. Different SPM scanning modes can be distinguished by the forces that govern the tip-sample interaction and by the tip-sample distance over which the forces operate. For AFM, there are three possible operational modes: the contact, intermittent contact (tapping) and the non-contact mode. In Fig. 2-14, these modes are related to the corresponding tip-sample distances and resulting potential variations.

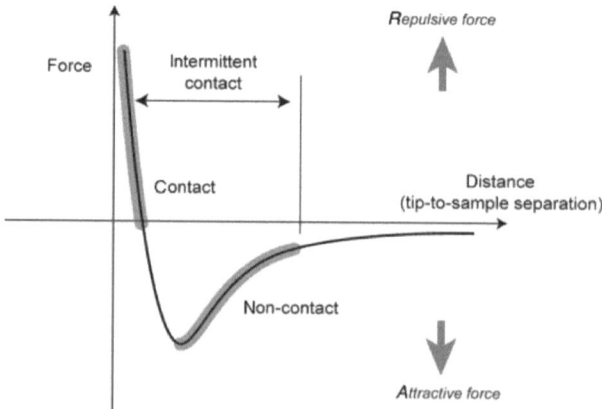

Fig. 2-14: Force-distance curve illustrating the interaction between an SPM tip and the surface of a sample [124].

The potential shown in Fig. 2-14 is determined by attractive and repulsive forces [123]. Van der Waals forces dominate the interaction in the attractive regime. These forces originate in the electromagnetic interaction of fluctuating dipoles, i.e. fluctuations of electron charge densities, located at the atoms of the tip and the surface. This phenomenon leads to induced dipoles in other atoms in the vicinity and the interaction is called *London force* [125]. The interaction of permanent dipoles with dipoles which are induced by permanent dipoles is called *Debye force* [126, 127]. The orientation force, or *Keesome force* [128], describes, in turn, the mutual interaction of permanent dipoles.

In summary, the van der Waals interaction can be expressed by a potential of the form

$$V(r) = -\frac{C_6}{r^6} \qquad (2\text{-}45)$$

where r denotes the tip-sample distance and C_6 the interaction constant, defined by London [125]. In the repulsive regime, strong short-range forces are present that impede further approach of the tip towards the sample.

Hamaker calculated the total tip-sample interaction by integration of the interaction potential between two macroscopic bodies [129]. By introduction of the *Hamaker constant*, which characterizes the resonance interactions between electronic orbitals in two particles and the intervening medium, analytical expressions for varying tip geometries could be derived. The Hamaker constant for the general interaction is ($\rho_{1/2}$ being the number of atoms per volume):

$$H = \pi^2 C_6 \rho_1 \rho_2. \qquad (2\text{-}46)$$

For instance, the force between the surface and a tip of spherical geometry with radius R at a distance D, is given by [123]:

$$F(D) = \frac{2HR^3}{3D^2(D+2R)^2}. \qquad (2\text{-}47)$$

A common approximation of the total potential that has to be considered for the tip-sample interaction is the *Lennard-Jones potential* [130]:

$$V(r) = \frac{A}{r^{12}} - \frac{B}{r^6}. \qquad (2\text{-}48)$$

For spherical tip geometry, the *Derjaguin–Muller–Toporov model* [131] yields a formula for the interaction force in the attractive and repulsive regime, taking into account both the elastic properties of tip and sample [132]:

$$F(z) = \begin{cases} -HR/\left[6(z_s+z)^2\right] & D \geq a_0 \\ -HR/6a_0^2 + \dfrac{4}{3}E^*\sqrt{R}(a_0-z_s-z)^{3/2} & D < a_0 \end{cases} \quad (2\text{-}49)$$

Here, z denotes the tip deflection, z_s the distance between the sample and the undeflected cantilever and E^* the effective contact stiffness, calculated from the respective elastic moduli and Poisson ratios of both materials.

Beside other forces as retarding effects, image forces and work function anisotropies, capillary forces are of particular importance under ambient air conditions. These forces are due to the presence of a water meniscus that forms with a large radius upon approaching the tip towards the surface. Discontinuous behavior in the tip-sample interaction can then result from the meniscus deformation that accompanies the variation in the tip-sample distance.

In this work, contact-mode and tapping-mode AFM (CM-AFM and TM-AFM, respectively) were employed. Contact mode AFM operates by scanning a tip attached to the end of a cantilever across the surface as shown in Fig. 2-15.

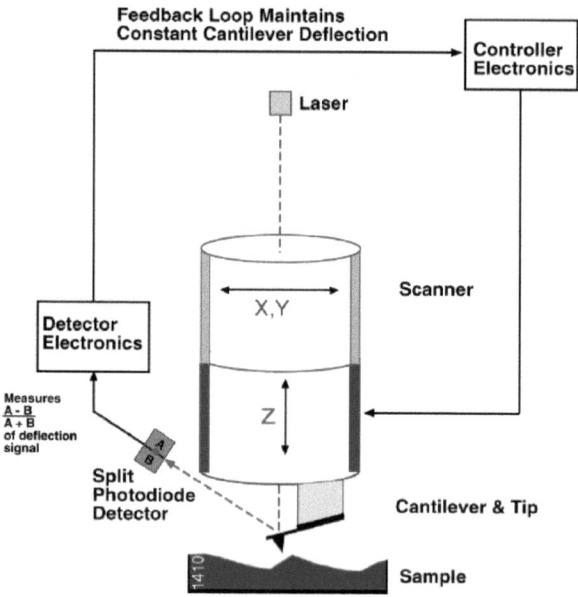

Fig. 2-15: Schematic of an AFM setup for contact-mode measurements [124].

The change in deflection of the cantilever is transformed to a change in the reflectance behavior of a laser beam and finally monitored by a photodiode detector. A feedback loop maintains almost constant deflection between the cantilever and the sample by vertical movement of the scanner at each (x,y) data point which preserves a pre-defined "setpoint" deflection value. By constant cantilever deflection, the force between the tip and the sample remains constant as well. Tapping-mode AFM operates by scanning a tip attached to the end of an oscillating cantilever across the sample surface. The cantilever oscillates at or near the resonance frequency with amplitude typically in the range of 200-400 kHz. The tip slightly "taps" on the sample surface during the scanning process and the feedback loop ensures constant amplitude. The vertical position of the scanner at each (x,y) data point which maintains a constant "setpoint" amplitude is stored by the computer to form the topographic image of the sample. Applying constant oscillation amplitude, a constant tip-sample interaction is preserved during imaging.

2.6 Scanning electron microscopy

High spatial resolution and high information depth of the scanning electron microscope (SEM) made it the primary choice for fracture analysis over the past decades [133]. Correspondingly, the investigation of fractally corroded silicon photoelectrodes, to be described in section 3.3, benefits from these exceptional capabilities and the instrument is introduced in the following in more detail. Max Knoll was the first to produce an SEM image in 1935 [134] followed by Manfred von Ardenne who worked on the physical principles of SEM imaging and the beam-sample interactions [135, 136]. The first commercial instrument was available in 1965, manufactured by the Cambridge Instrument Company under the name "Stereoscan".

2.6.1 Experimental arrangement

SEM employs a beam of electrons accelerated towards the surface of a specimen. Electrons, escaping from the sample after single or multiple scattering processes, are then detected. Usually, the electron beam is generated by thermionic emission from a heated tungsten filament (F) as shown in the principle drawing in Fig.2-16.

If higher brightness, i.e. number of electrons per second and unit area, and lower energy spread (~0.5 eV) is required, sharp tips are used for field emission (see insets in Fig. 2-16).

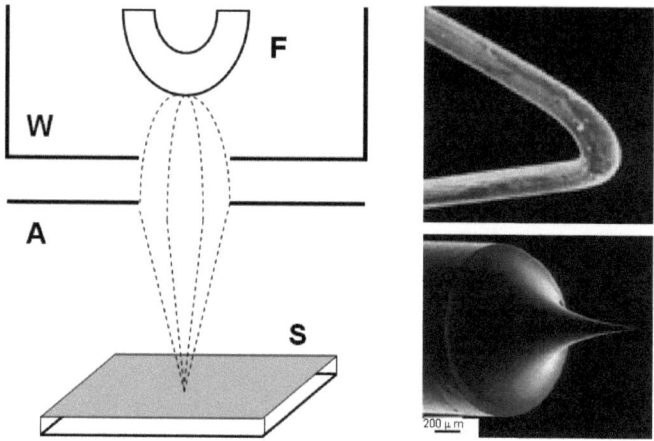

Fig. 2-16: Left: electron emission by thermionic emission from a tungsten filament (F). A potential difference between the filament and an anode (A) accelerates the electrons towards the sample (S). A biased Wehnelt cylinder serves for spot size adjustment. Right: transmission electron microscopy images of a tungsten filament (above) and a sharp tungsten tip (below) for field emission microscopy [133].

The electrons are accelerated by a potential difference (F-A) towards the sample to generate a beam of controlled energy. Electronic lenses such as a Wehnelt cap (W), set to a slightly more negative potential than the filament, facilitate beam focusing. The area of the filament from which electrons are emitted can thereby be limited to a smaller spot. Primary electrons, emitted by the filament (tip), enter the specimen and are scattered depending on the probability of a single scattering process. This process can be expressed either by the scattering cross-section σ, i.e. the apparent area which the scattering particle presents to the electron, or the mean-free path, λ, which denotes the average distance an electron can travel before being scattered. Both magnitudes are mutually related by:

$$\lambda = 1/N\sigma. \tag{2-50}$$

N denotes here the number of scattering particles per unit volume. Scattering of an electron in a thick specimen can occur many times (multiple scattering effects). The corresponding probability distribution goes beyond simple mathematical treatment by, e.g. the Poisson distribution, and modern analysis applies Monte Carlo simulation to account for all the processes involved. So far, only elastic scattering was considered, i.e. the electrons deflected by the specimen do not experience a loss of energy. This type of scattering, also known as *Rutherford scattering*, is due to the electrostatic interaction of the electron with the nuclei and all electrons of the scattering particles during its passage through the material.

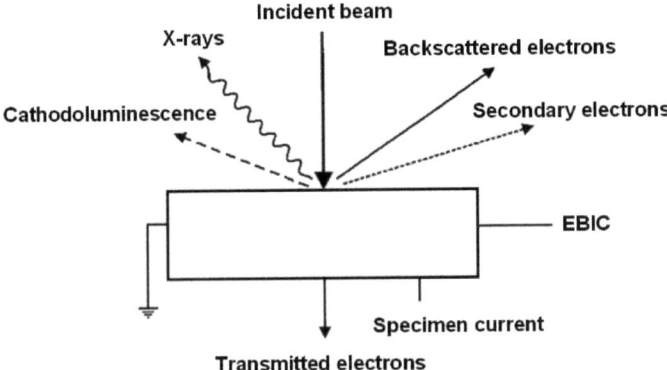

Fig. 2-17: Measurable quantities in scanning electron microscopy after excitation with an electron beam of controlled energy. The abbreviation EBIC stands for electron beam induced current.

Secondary effects are characterized by a detectable amount of energy loss (of the order of 0.1 eV and above). There is a large number of possible inelastic scattering processes such as phonon or plasmon scattering that can lead to emission of X-ray radiation, luminescent light or secondary electrons which all can be used for both imaging and analysis. The number of secondary/backscattered electrons for each incident electron is expressed in the secondary/backscattered electron coefficient. A survey of detectable signals is shown in Fig. 2-17, including electron beam induced currents which can be analyzed for the detection of electrostatic fields present across the sample.

The SEM operates in a scanning mode, i.e. the movement of the electron beam across the surface is related to the movement of a corresponding electron beam in a cathodic ray tube (CRT), as part of a monitoring system. This technical principle is depicted in Fig. 2-18 where the imaging system is shown comprising condenser and objective lenses, generating a 2-10 nm wide electron spot, and the scan coils which ensure the controlled horizontal and vertical beam movement. The scan generator couples the respective positions of the probing and the imaging beam. Additionally, the variation of the signal with changing beam position can be evaluated by a waveform monitor.

The secondary electron signal is the most commonly used in scanning electron microscopy. The term 'secondary electrons' is not well defined but refers in general to electrons of kinetic energies below 50 eV. The electron detection is realized by, e.g., Everhart-Thornley detectors which combine a scintillator, biased to an appropriate acceleration voltage to ensure the excitation by slow electrons, and a photomultiplier unit.

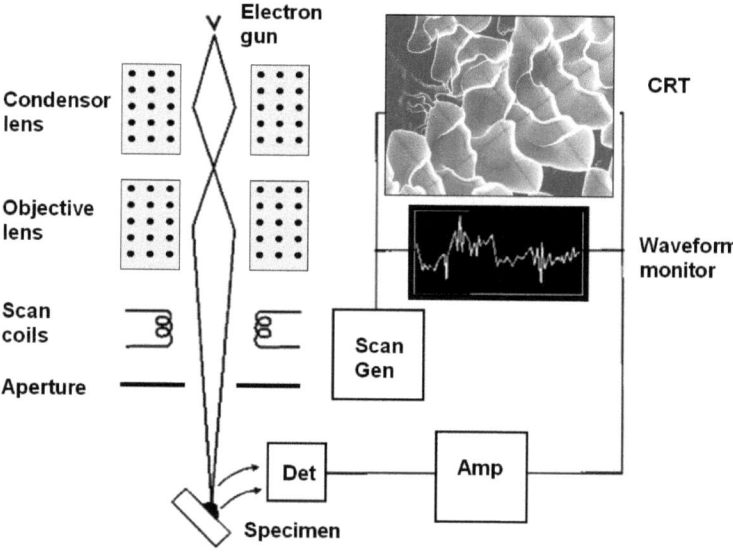

Fig. 2-18: Schematic set-up of a scanning electron microscope. The beam of an electron gun is focused towards a specimen. Resulting signals such as backscattered or secondary electrons are detected, amplified and mapped by a monitor system. Pairs of scan coils allow manipulation of the electron beam position in x- and y-direction.

In principle, the same detection device can be used for backscattered electrons which have higher energies but are less numerous. If the scintillator bias is switched off, only high energetic electrons will be detected, secondary electrons will be excluded. In high resolution SEM (HR-SEM), so-called through-the-lens detectors are used. Here, the microscope has specially designed objective lenses with large magnetic fields and low spherical aberration. By placing a scintillator within the lens system, the microscope has very good electron collection efficiency and operates at short working distances.

2.6.2 Depth of field, chemical and spatial resolution

For interpretation of SEM images, it is important to know which sample properties constitute the contrast. The electron yield of backscattered electrons increases with atomic number while the corresponding yield of secondary electrons varies little (see Fig. 2-19). Therefore, in an image stemming from backscattered electron, chemical information is superimposed on height information.

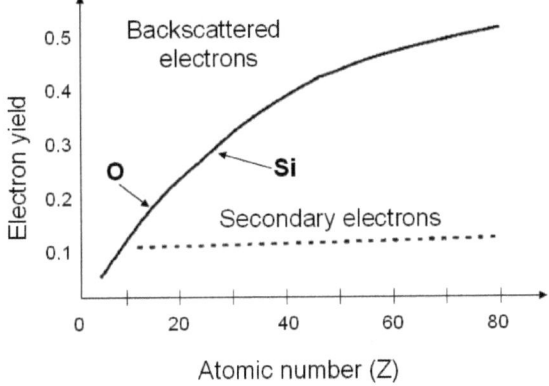

Fig. 2-19: Electron yield for backscattered and secondary electrons in dependence on the atomic number. The yield for silicon (A = 28) and oxygen (A = 16), respectively, are indicated by arrows [133].

The best spatial resolution is achievable by secondary electrons which have their maximum intensity in a direction along the incident beam. The signal originates from a small detection volume that is little larger than the so-called sampling volume, i.e. the volume reached by the incident electrons.

The sampling volume for different SEM signals is shown in Fig. 2-20. For instance, X-ray radiation originates from deeper sample areas than secondary electrons. The backscattered electron coefficient also shows dependence on the crystallographic orientation of the sample with respect to the incident beam. This effect results from the diffraction of the incident beam in dependence on the depth of penetration and can be understood as a channeling effect that makes it less probable for backscattered electrons to escape the farther the electrons are able to penetrate the specimen.

The spatial resolution achieved by specific microscope adjustments is sufficient to observe grain boundaries of ~ 100 nm size on otherwise smooth topographies.

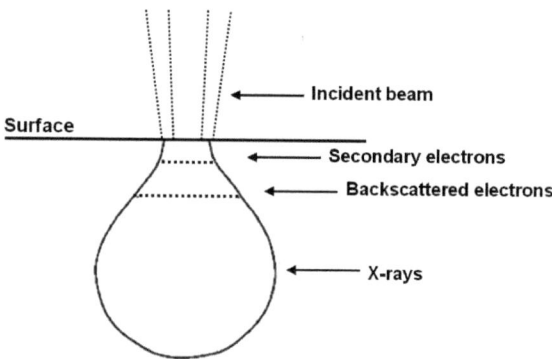

Fig. 2-20: Sampling volumes for the measurable quantities after excitation with an electron beam of controlled energy.

Mathematically, the high spatial resolution of SEM can be assessed by consideration of the small *de Broglie wavelength* of the incident electrons. If these electrons are accelerated by the potential V, the kinetic energy can be expressed by $E_{kin} = eV$, where e 1.6 x 10^{-19} C denotes the electron charge. With momentum $p = mv$ and Planck constant h the de Broglie relationship reads:

$$\lambda = \frac{h}{p} = \frac{h}{mv}. \tag{2-51}$$

The electron mass is determined by

$$m = \frac{m_e}{\sqrt{1-(v/c)^2}} \tag{2-52}$$

where m_e = 9 x 10^{-31} kg is the electron rest mass. With the energy-mass-relation $E_{kin} = eV = (m - m_e)c^2$, c being the speed of light, it follows from Eqs. 2-51 and 2-52:

$$\lambda^2 = \frac{h^2}{\left(2eVm_e + (eV/c)^2\right)}. \tag{2-53}$$

Substituting e, c and h by the corresponding values, the wavelength can be expressed as function of the acceleration potential:

$$\lambda = \sqrt{\frac{1.5}{(V + 10^{-6}V^2)}} \; nm. \tag{2-54}$$

Eq. 2-54 holds for lower kinetic energies in SEM, where relativistic corrections are negligible, as well as for higher energies in TEM ($E_{kin} > 2 \times 10^4$ V). A simplified formula of Eq. 2-54 is often used in SEM: $\lambda = \sqrt{\dfrac{1.5}{V}}$ nm.

In order to understand the high depth of field in SEM, exemplifying calculations are outlined in the following: the relation between probing area of the beam, i.e. the electron spot size, and the detecting monitor screen is dependent on the pixel size that is given by the monitor. If the monitor screen can be divided in, e.g., 100 x 100 pixels, each of them having a size of 100 μm, then the corresponding pixel size on the specimen is given by

$$p = \frac{100}{M} \mu m \tag{2-55}$$

where M denotes the magnification of the microscope. In order to optimize both spatial resolution and electron yield, the corresponding pixel sizes of the specimen and the screen should correspond to each other. If the electron spot is smaller than p, the electron collection will be reduced. If the spot is larger, a screen pixel obtains information from neighboring pixels, thus reducing the spatial resolution.

The depth of field can now be understood assuming the quantities given above. The electron beam is focused on the sample with convergence angle α. This is illustrated in Fig. 2-21 where the determining quantities as aperture, working distance and convergence angle are indicated. The distance h over which the specimen will remain in focus, i.e. the depth of field, is given by

$$h = \frac{100}{\alpha M} \mu m. \tag{2-56}$$

Considering the relation between convergence angle, aperture of the objective lens and working distance, h can be finally expressed as

$$h = \frac{200 WD}{AM} \mu m. \tag{2-57}$$

An approximate calculation (M = 1000, WD = 20 mm, α = 0.7) shows that the depth of field is ~ 40 μm in SEM while the corresponding value in optical microscopy is reduced to ~ 1 μm.

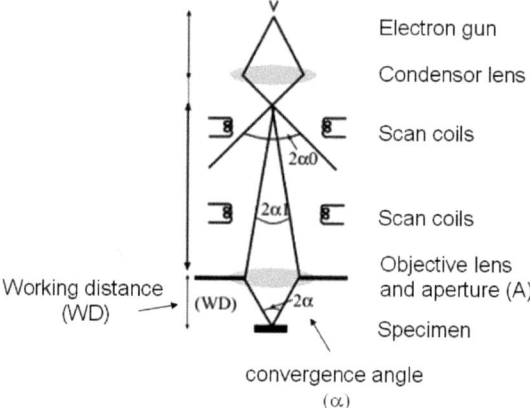

Fig. 2-21: Diagram of the beam pathway in SEM and resolution limiting quantities as convergence angle α, working distance WD and aperture A.

3. Results and discussion

The formation of micro- and nanotopographies on Si(111) during chemical and (photo)electrochemical conditioning will be described in the following chapter. It is attempted to identify the most determinant factors for the structure shape and structure distribution. Model development and computer simulations are presented. Variation of the Si(111) miscut angle as well as the use of surface orientations such as (100) and (110) are employed to clarify the influence of dissolution anisotropies. Brewster-angle reflectometry as an appropriate optical *in situ* technique is introduced allowing real-time assessment of the topographical state of the samples and real-time control of the external electrochemical parameters, potential and light intensity. The chapter begins with the investigation of successively etched native oxides on Si(111) as test structure for the sensitivity of the technique. An optimized H-termination procedure with regular surface topography could thereby be developed. Etch pit initiation in NH$_4$F solution is related, in contrast to other authors, to the underetching of accumulated reaction products at the Si interface.

3.1 Identification of a sub-surface stressed silicon layer

3.1.1 Introductory remarks

Since the work of Higashi et al. [137, 138], etching of Si(111) in concentrated NH$_4$F (40%) was intensively analyzed in numerous publications. The perspective of a well-defined smooth surface topography prompted investigations of the reaction mechanism [57, 139], H-termination and stability [138, 140-144], miscut angle dependence [145], influence of solution concentration and composition [146-148] and finally sophisticated model development and computer simulations [149-151]. The stepped (111)-surface contrasts with findings on Si(100) where typically roughness on the atomic level is observed after HF-treatment [152]. The relation of the surface morphology to initial processes as during surface oxidation or metal deposition is therefore best investigated on the Si(111) surface [153-156]. In turn, the outstanding anisotropic properties of the Si(111)-NH$_4$F system led to uncovering of rudimentary anisotropies of the Si-HF system and smoothening procedures of Si(100) by hot NH$_4$F solutions [157, 158]. Compared to the extensive studies of structural, chemical and kinetic phenomena of Si etching in NH$_4$F, the role of the SiO$_2$ layer and its interface for the evolution of the regular Si(111) topography attracted less attention [159, 160]. Since the interfacial SiO$_2$/Si region is characterized by strained atomic bonds at and beneath the interface, an influence on the formation of terraced topographies appears very likely. Interfacial strain and stress effects were addressed in several publications of the past years [6, 7, 32-34, 41, 161, 162] and it is now unquestioned that, beside substoichiometric oxides at the interface, a bulk-near transition layer of strained or stressed silicon increases the nominal thickness of the whole silicon oxide/silicon system by a few additional nm. However, while structure and stoichiometry of the interfacial region are still discussed, even for the frequently investigated $SiO_2/Si(100)$ system, quantitative assessment of the stressed region is still challenging and little is known about the effects onto the etching process in wet chemical or electrochemical preparations.

In the following, the smooth etching behavior of Si(111) in concentrated ammonium fluoride is exploited in order to assess the dissolution of native and thermal oxides until the interfacial region is reached. The sensitive optical response of BAA to minute topographic changes at the surface is evaluated in order to characterize the dissolution process in detail. By this approach, it is possible to relate an unusual surface roughening in the etching process to the presence of the stressed interfacial region. A transitory reflectance behavior, observed during *in situ* BAR monitoring, is furthermore attributed to accelerated interface dissolution. Considering these results, an optimized preparation method of chemically clean Si(111) with regular atomic steps was developed and is finally presented.

3.1.2 *Ex situ* Brewster-angle analysis: in loco etching results

Si(111) samples were prepared from Cz-grown phosphorous doped n-type silicon (Sico GmbH, Germany), $N_D = 10^{15}$ cm^{-3}, with nominal 0° miscut angle. The samples were stored in a laboratory environment for saturation of native oxide growth and were cut (1.5 cm x 1.5 cm) with defined azimuthal orientation towards the so-called primary flat. The primary flat represents (for Si(100) and Si(111)) a standard edge on the respective wafer disks, indicating the <110> direction. Mounting to the sample holder of the Brewster-angle spectroscope was realized such that the respective solutions could be approached from beneath without changing the position of the sample (*in loco* cleaning and etching).

The pre-treatment of the samples consisted of: (i) water rinse (Milli-Q water, 18.2 MΩ, 1 min.), (ii) ethanol rinse for 1 min (p.a. grade, Merck); the procedure was repeated once followed by drying in high-purity (6N) nitrogen stream. For etching, 40% NH_4F (ultrapure, Merck), *p*H ~ 7.5, was used with 20 s exposure time per step followed by a 1 min water rinse and N_2 drying. In Fig.3-1, the BAA results for cleaning and subsequent etch-back steps of the native oxide of silicon are shown. The data were obtained at 500 nm wavelength corresponding to a photon energy of 2.48 eV. The effects of surface cleaning are visible by the large signal change from situation A to B. These changes are larger than for the successive etch steps which are shown until inflection of $R_p(\varphi_B)$ and of φ_B occurs. It can be seen that, in general, the changes in the reflectivity are larger than those of the Brewster angle. After four steps, a minimum of $R_p(\varphi_B) = 3.7 \times 10^{-4}$ is reached. The corresponding change of the Brewster angle results in a maximum of $\varphi_B = 76.895°$.

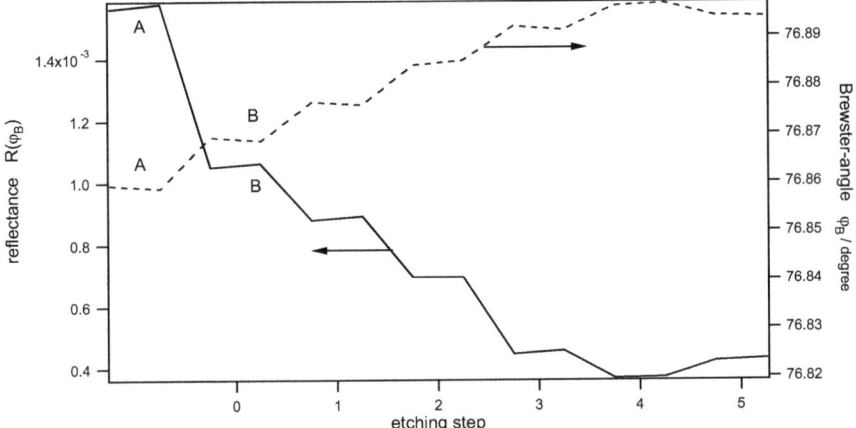

Fig. 3-1: Measured Brewster angle φ_B and reflectance $R_p(\varphi_B)$ at this angle, for cleaning and five etch steps in NH$_4$F (40%). (A) Surface 'as-cut'. (B) After ethanol and ultra pure water cleaning.

Fig. 3-2 shows contact mode AFM images after the first (Fig. 3-2a) and the fifth (Fig. 3-2b) etching step. In Fig. 3-2a, the original experimental data are displayed in the left half of the picture; the right half shows an image obtained after mathematical processing using an autocorrelation procedure that facilitates the detection of weak surface patterns. By this procedure, a convolution integral across the surface area is calculated that measures the average distance of surface sites with similar height properties. The integral is exemplified for the one-dimensional case below:

$$ACF(l) = \left(\frac{1}{L}\int_0^L z(x) \cdot z(x+l) dx \right) / \sigma^2. \quad (3\text{-}1)$$

Here, $z(l)$ denotes the surface height at position l while x represents the shift to be employed until another point $z(l')$ of equal height is reached. The integral is normalized with respect to the total length L and the root mean square roughness σ. The two-dimensional integral can be expressed accordingly.

Whereas the AFM height-image (Fig. 3-2) appears rather unstructured, the autocorrelation image reveals a terraced surface topography. After etch step 5 (Fig. 3-2b), this topography occurs also in the CM-AFM image. It should be noted that parallel aligned terraces are visible with identical terrace widths in both figures. The root mean square surface roughness (rms), monitored after each step by AFM, did not change significantly (Fig.3-2a and b).

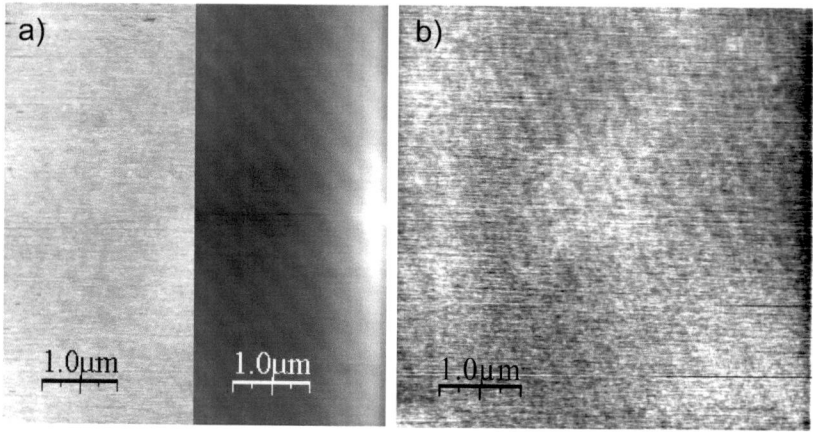

Fig. 3-2: (a) AFM image (left half) and corresponding autocorrelation image (right half) after the first etching step; the two-dimensional autocorrelation function provides enhanced perception of surface patterns, revealing thus the presence of terraces already after the first etching step. (b) AFM image of the nearly oxide-free surface after five etch steps.

Fig 3-3 shows the development of the BAA signals for continued etching.

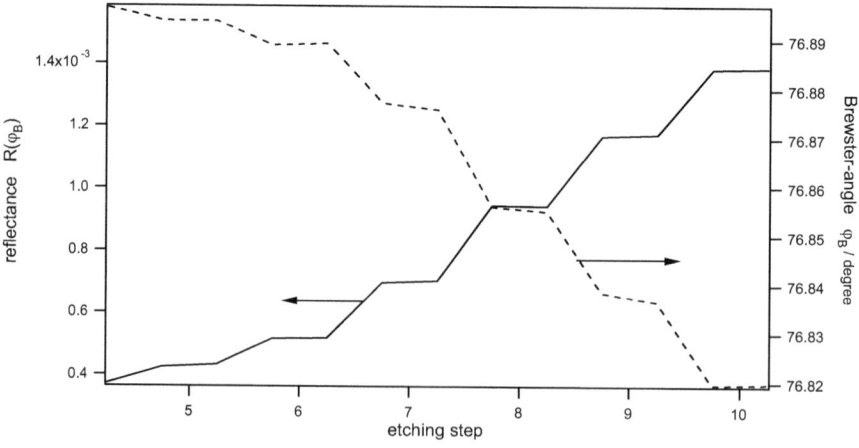

Fig. 3-3: Brewster angle φ_B and reflectance $R_p(\varphi_B)$ at this angle, for etch steps four to ten.

68

A distinct reflectance increase by a factor of ~3 and a simultaneous decrease of the Brewster angle is observed. It can be seen that the decrease of φ_B proceeds much faster than its increase during the initial etch sequence (Fig. 3-1) between steps 1 and 4. Also, the increase of R_p appears more pronounced compared to the changes of the first four etching steps. The final value of R_p corresponds to a value between A and B in Fig. 3-1 whereas the value of φ_B (for etch step 10) is considerably smaller (76.82°) than the initial Brewster angle (76.86°) before processing of the sample.

The corresponding AFM images taken after etch step 6 (Fig. 3-4a) and 8 (Fig. 3-4b) show a pronounced increase in surface roughness for only two successive etch steps. This is illustrated by the two-dimensional images and cross sectional analyses in Fig. 3-4. The calculated rms values obtained by two-dimensional analysis of 1 µm x 1 µm areas change approximately by 0.15 nm for each etch step.

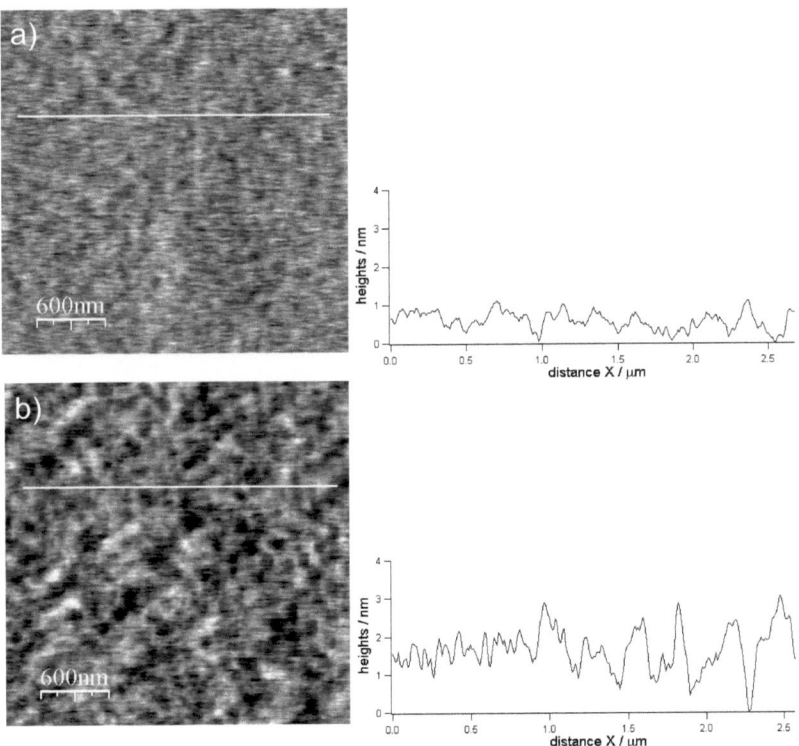

Fig. 3-4: AFM data and cross-sections. (a) For etch steps 6. (b) For etch step 8 (see text).

These findings make clear that optical analysis, to be carried out below, has to consider roughness layers on top of the silicon sample and, respectively, at the SiO$_2$/Si interface. Model calculations will be based on Maxwell-Garnett effective medium approximation (EMA) [101] which is applicable for small surface corrugation until etching step 5 as concluded from the respective AFM analyses (see section 2.1.2):

$$\frac{\langle\varepsilon\rangle-\varepsilon_h}{\langle\varepsilon\rangle+2\varepsilon_h} = v_1 \frac{\varepsilon_1-\varepsilon_h}{\varepsilon_1+2\varepsilon_h} + v_1 \frac{\varepsilon_2-\varepsilon_h}{\varepsilon_2+2\varepsilon_h} + \dots . \tag{3-2}$$

In this formula, $\langle\varepsilon\rangle$, ε_h, ε_1, ε_2,..., denote the complex dielectric functions of the effective medium, the host medium and inclusions of type 1,2,..., with volume fractions v_1, v_2,..., of material 1,2,..., in the considered volume. Roughness and surface profiles determined by AFM provide therefore information about the appropriate fraction to be used in Eq. 3-2.

The unexpected roughening effect observed after repeated exposure of the sample to the NH$_4$F solution was analyzed by further experiments where samples after continued etching were analyzed with AFM. For that purpose, etching times larger than two minutes after 100 s pre-etching time (corresponding to etch step 5 in Fig. 3-1), as indicated on the horizontal axis in Fig. 3-5, were chosen and the resulting rms values were determined. Interestingly, although R$_p$(φ_B) (open circles) changes considerably, the rms roughness shows little change, providing first indication of a sub-surface layer that influences the optical data. For quantitative assessment of the chemical state of the surface at different times, SPRES measurements were performed on samples prepared in separate experiments but under comparable conditions. During transfer to the vacuum chamber, samples were stored in a N$_2$-

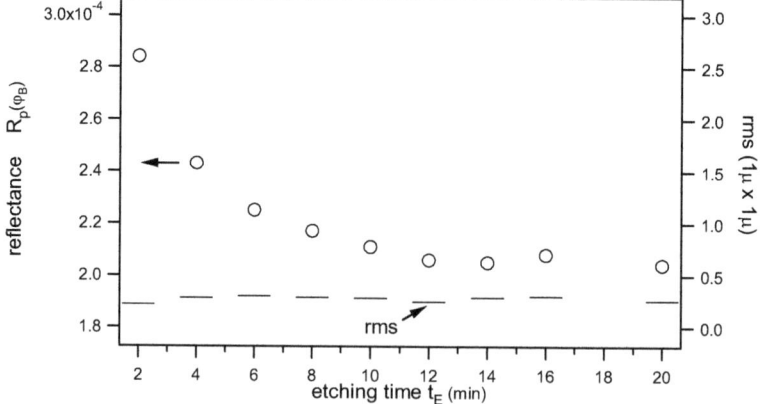

Fig. 3-5: (a) Extended etching of Si(111) after an initial etching step of 100 s; experimental reflectance values (open circles) show minor fluctuations from t_E = 10 min on. Corresponding CM-AFM rms values (lines) show throughout little variation.

filled glass tube. Contamination levels by hydrocarbons were thereby kept below 0.2 monolayer overall coverage according to quantitative analysis of the SRPES data. In Fig. 3-6 and 3-7, SRPES results for the Si $2p_{1/2-3/2}$ core level at a photon energy of hv = 170 eV are shown. At the corresponding kinetic energy of excited photoelectrons of ~70 eV, the mean inelastic scattering length λ is very short, resulting in ultrahigh surface sensitivity in the detection of elastic photoelectrons (λ ~ 0.4 nm). In Fig. 3-6, the spectrum for condition B, related to the unetched native oxide layer, and for successive etch steps are shown. A decrease of the signal around E_{kin} = 66.5 eV and a corresponding increase of the core-level signal around 70 eV can be observed. For etch step 8, the signal at 66.5 eV which has to be attributed to SiO_2 is very small.

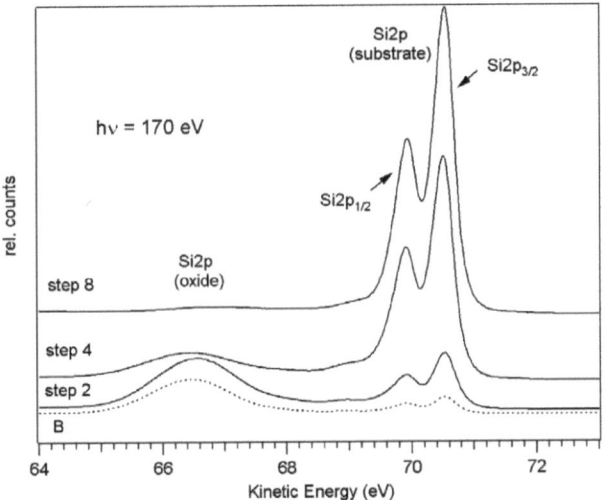

Fig. 3-6: Si 2p SRPES data at photon energy hv = 170 eV for the sample after cutting and cleaning ('B' in Fig. 3-1, dashed line) and for ecth steps 2, 4 and 8 (full line).

The distribution of substoichiometric silicon oxides was analyzed by deconvolution of the integral Si 2p signal for etch steps 4 and 8 as shown in Fig. 3-7. The pronounced contributions from substoichiometric oxides to the signal suggest the presence of silicon in lower oxidation states preferentially near the interface region. A more detailed analysis will follow below in section 3.1.3.

The correlation of optical and SRPES data necessitates the knowledge of optical reference values of smooth H-terminated Si(111) as obtained in well defined experiments. These reference values were taken from published ellipsometry experiments performed in inert gas atmosphere at a fixed angle of incidence that was several degrees off the Brewster angle [95].

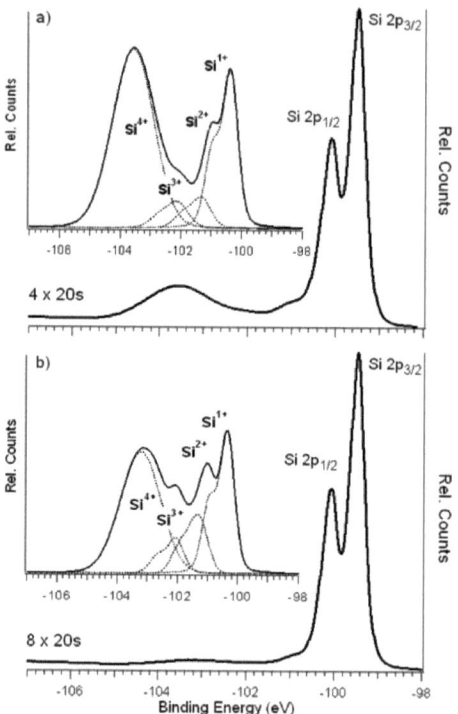

Fig. 3-7: SRPES results, obtained after etching times of 80 s and 160 s of native oxide covered Si(111). The excitation energy was hν = 170 eV. In a) and b), the deconvolution of the Si 2p envelope shows the respective contributions of different oxidation states to the photoelectron signal.

According to the mathematical relation between the dielectric function and the reflectance behavior of solids, described in section 2.1, a reflectance parabola was calculated for a nearly ideal H-terminated Si(111) surface as corresponding set of BAA reference values. The calculation was performed based on the assumption that the Brewster angle and the reflectance at this angle can be analytically related to each other. The validity of this assumption is restricted by two effects: firstly, published values were provided as components of the dielectric function, i.e. *after* calculation of the originally measured ellipsometric angles Δ and Ψ [91, 97]; secondly, Brewster-angle and reflectance for ideal H-terminated surface were calculated according to a two-layer model (ambient / substrate) without consideration of roughness or contamination. A justification of this approach will be given further below by comparison of data obtained by ellipsometry and Brewster-angle analysis for an optimized surface condition on H-terminated Si(111).

Using these BAA reference values, multi-layer analysis was applied to the experimental values, measured after various etching times. As an example for the calculation procedure, measured $R_p(\varphi_B)$ values (open circles) are contrasted in Fig. 3-8 with those calculated from the reference values. For the calculation (dashed line), AFM-rms data as measured at etching time t_E = 2 min were used in a three-layer model comprising ambient, a surface roughness layer and silicon bulk.

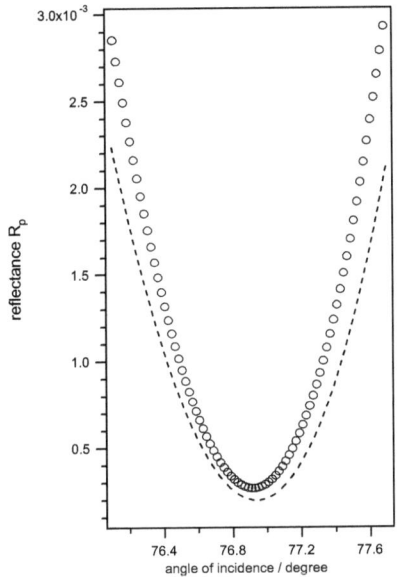

Fig. 3-8: BAA analysis after 2 min etching: experimental values (open circles) compared to simulated data (dashed line) according to an ambient / silicon surface roughness / silicon bulk three-layer model only (without assumption of a strained layer).

A distinct difference can be noted between calculated and measured $R_p(\varphi_B)$ values. Fig. 3-9 shows this difference for all extended etching times between 2 min and 20 min after the minimum of $R_p(\varphi_B)$ in Figs. 3-1 has been reached (etch step 5, corresponding to 100 s pre-etching time). Here, the measured reflectance at the Brewster angle is opposed to the corresponding calculated reflectance and indicated as difference value $\Delta R_p(\varphi_B)$. The calculation was performed again on the basis of a three-layer model (ambient / roughness as determined by AFM / silicon bulk) using silicon bulk data from ellipsometry measurements. It can be clearly seen that the assumption of the three-layer model is not sufficient to describe the surface/interface condition of the samples until $t_E \sim 10$ min and confirms the assumption of a sub-surface layer, firstly discussed with the results in Fig. 3-5. At $t_E \sim 10$ min, $\Delta R_p(\varphi_B)$ in Fig. 3-9 vanishes. At this condition, the corresponding CM-AFM image (Fig. 3-10) shows large smooth atomic terraces, indicating a high quality H-terminated surface. This surface is typical for prolonged etching in NH$_4$F producing (1x1) H:Si(111) surfaces [138]. The actual miscut angle of the samples was calculated to point $\sim 0.05°$ towards the $<11\bar{2}>$ direction. Comparison with Fig. 3.2 shows furthermore good agreement with the surface topography observed already after the first etch step.

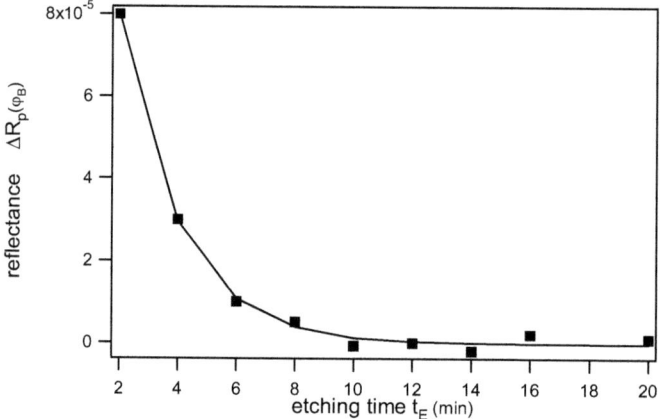

Fig. 3-9: Difference reflectance $\Delta R_p(\varphi_B)$ for continued etching time t_E (squares). $\Delta R_p(\varphi_B)$ is obtained by the difference between experimental data (see Fig. 3-5) and theoretical data for an ambient / silicon surface roughness / silicon bulk model. An exponential-like decrease (fitting curve) is visible. At $t_E \sim 10$ min, a three-layer model is sufficient to model the experimental data.

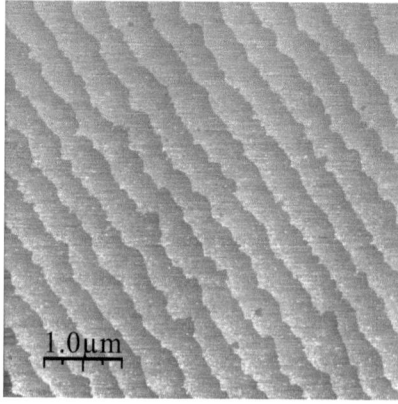

Fig. 3-10: CM-AFM image after 10 min etching of the oxide-free Si surface. The terraces are regularly shaped; very few etch pits are visible.

A comparison of measured reflectance data and those expected from ellipsometry values show very good agreement after etching times $t_E \sim 10$ min as depicted in Fig. 3-11. In this case, the assumption of a three-layer model (ambient / roughness as determined by AFM / silicon bulk) is sufficient to simulate the corresponding data. In Fig. 3-10, theoretical data (calculated from the reference parabola) and experimental data are compared for 10 min etching.

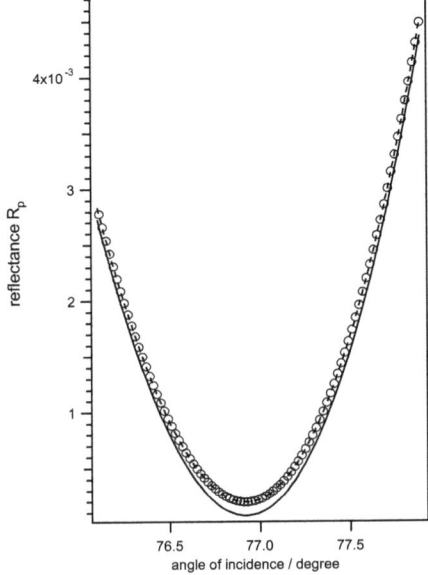

Fig. 3-11: BAA analysis of H-terminated Si(111) after 10 min etching of the oxide-free Si surface: published Si bulk data obtained by ellipsometry (full line) were used to calculate from experimental values (open circles) the simulating parabola (dashed line). A three-layer model describing an ambient/ surface roughness / bulk system was assumed.

In addressing the question of the discrepancy between $R_p(\varphi_B)$ data, mathematically obtained by a three-layer model and shown in Figs. 3-8 and 3-9, one has to consider the respective surface sensitivity of the methods: AFM monitors the top surface topography whereas the BAA signal probes the interface region within the penetration depth of the probing light. It can be assumed that, for etching times $t_E < 10$ min, the deviation between calculated and experimental values results from the sub-surface region of the silicon sample. The existence of a strained Si layer located between silicon dioxide and the undistorted bulk was addressed in section 1.1.3. The characterization of the optical properties of this layer, obviously present in our experiment, is a perquisite for accurate evaluation of the optical data.

In an additional experiment, the oxide was therefore removed by etching for 100 s in 40% NH$_4$F (see Fig. 3-1, etching step 5). Then, the rms roughness of the interface between the strained layer and the oxide was determined. Furthermore, the strained layer was etched off (t_E = 10 min in 40% NH_4F); the appropriate etching time was determined from results shown in Fig. 3-9 where $\Delta R_p(\varphi_B)$ approaches 0. The unknown layer thickness was chosen to be d = 4.5 nm according to published etch rates of 0.45 nm/min at 21.4° C for bulk silicon [163]. CM-AFM rms roughness values of the exposed surface were used for further calculation. Applying an effective medium approach and a least-square fit procedure of the Brewster-angle parabola (see appendix A.1), the optical constants of the strained layer were determined to ε = (16.80 ± 0.05) + i*(0.50 ±0.05). The representation of the determined multi-layer system is schematically shown in Fig. 3-12.

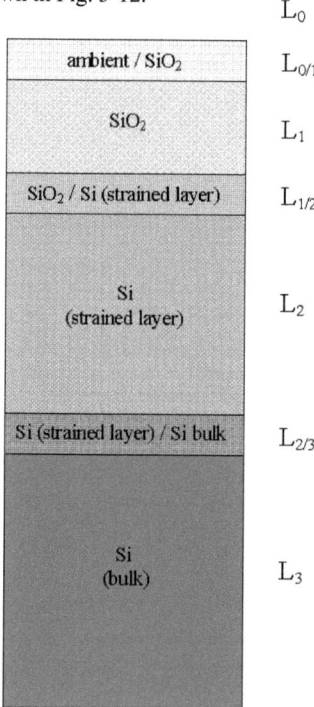

Fig. 3-12: Schematic representation of the determined multi-layer-stack. Surface and interface layers are indicated as L_0, L_1, L_2, L_3 and $L_{0/1}$, $L_{1/2}$, $L_{2/3}$, respectively. Thicknesses and void fractions of the interlayers $L_{2/3}$, $L_{1/2}$ and $L_{0/1}$, determined by AFM, were 0.32 nm/0.70, 0.41 nm/0.66 and 0.38 nm/0.69 (for the unetched sample). Oxide thickness evaluation (see Fig. 3-14) is based on this model.

The system consists of seven components and includes silicon oxide and its outer interface towards the ambient. Using this model, quantitative evaluation of the optical data during successive etch-back steps is now possible. The calculation was carried out in a three-step procedure: firstly, the parabola measured at position B in Fig. 3-3 was used to calculate the refractive index n of the oxide by taking into account the optical constants of the strained layer and of all respective interfacial regions. The value n = 1.55 obtained by this calculation is smaller than described by published curves of the $n(d_{ox})$ relationship of ultrathin thermal oxides [164, 165]. Secondly, upon etching, the measured parabolae, as shown in Fig. 3-13, were fitted using the calculated (n, d_{ox})-pairs. According to [164, 165] an increase of the refractive index was determined towards the silicon interface. The slope, however, is smaller with a maximum value of n = 1.80 at etch step 4. The figure shows clearly that using the optical constants of the strained layer and the interfacial roughnesses, based on surface topographic assessment by AFM, excellent agreement with experimental data is achieved.

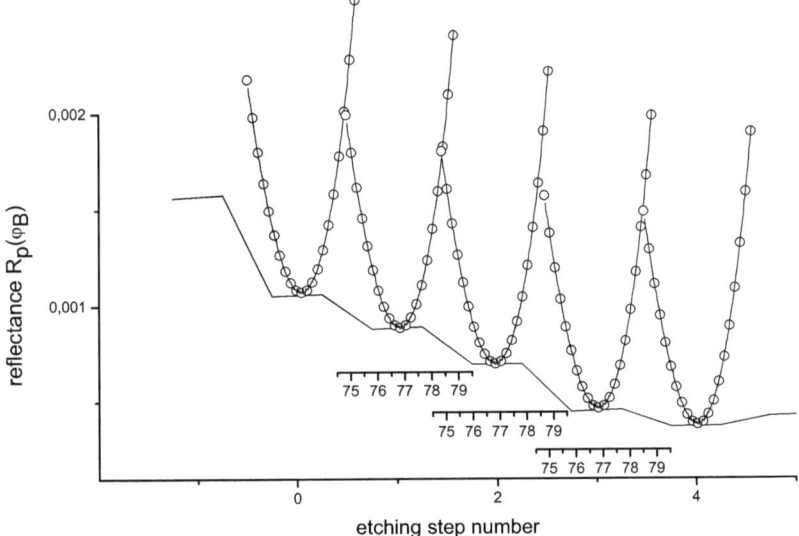

Fig. 3-13: BAA analysis for etch steps 0 (as cut and cleaned), 1, 2, 3 and 4. The corresponding experimental parabolae are shown as insets with a reduced number of measured points (open circles) and are compared to modeled data according to the seven layer model described by Fig. 3-12.

Subsequently, the oxide layer thickness was determined for etching steps 0 through 4 (see circles in Fig. 3-14). The Brewster-angle measurements indicate that the initial native oxide layer had a thickness of about 1.2 nm which corresponds to published values of as-received Si(100) samples which were exposed to air for several years [166]. With repeated etching steps, the change in oxide thickness shows some variation (between 1 Å and 5 Å per etching step). This observation may indicate beginning three-dimensional etching with etching step 2.

Finally, SRPES data, as shown in Fig. 3-7, were evaluated for the cleaned sample and three etching steps by analysis of the Si bulk signal compared to the total area of the Si 2p core-level shifts. With the intensity ratio of a bare silicon substrate with respect to an infinitely thick oxide layer, I_0/I_∞, and the escape depth of photoelectrons for the given excitation energy, λ_{SiO_2}, the integral oxide signal was calculated by [24]:

$$d_{SiO_2} = \lambda_{SiO_2} \ln(1 + \frac{I_{SiO_2}}{I_{Si}} \cdot \frac{I_0}{I_\infty}), \qquad (3\text{-}3)$$

The results are shown in Fig. 3-14 as full squares and indicate excellent agreement between optically determined oxide thicknesses and those obtained by SRPES. This agreement also confirms that minute changes of the chemical state at the silicon surface can be reliably monitored by Brewster-angle analysis. The resolution limit of the method is well below 1 Å.

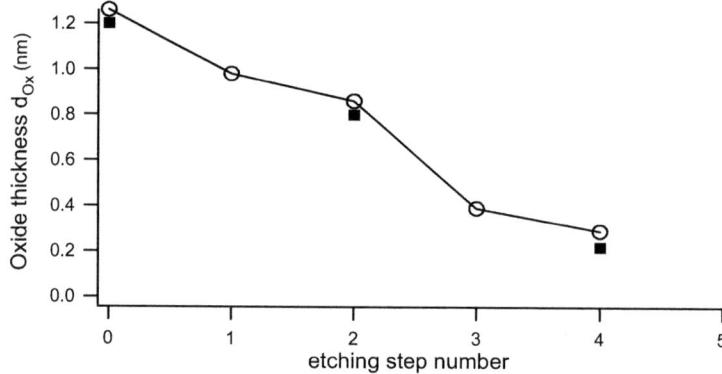

Fig. 3-14: Comparison of oxide thicknesses measured by BAA analysis (open circles) and SRPES (full squares)

The roughness increase for etch steps 6-10 as shown in Fig. 3-3 appears, at first glance, in contradiction to the well-known smoothing effect of the NH$_4$F treatment [138]. The data in Fig. 3-9 show, however, that after an overall etching time of 3 min (with 100 s pre-etching and intermediate UPW-rinse), the silicon surface is smooth and remains smooth. This effect was investigated by application of *in situ* BAR in the following section.

3.1.3 *In situ* Brewster-angle reflectometry: real-time monitoring results

In order to obtain further details about the unexpected roughening behavior during etch-back, native oxide covered Si(111) samples, as used in the preceding section, were immersed in

NH$_4$F containing solutions of varying concentrations and monitored by *in situ* Brewster-angle reflectometry at 500 nm wavelength. During sample alignment, de-ionized water was used, and the angle of incidence was adjusted to the Brewster-angle of the SiO$_2$-Si-system immersed in water. Then, water was exchanged for the etching solution.

During etching in 40% NH$_4$F, the reflectance signal $R_p(\varphi_B)$ decreases for 100 s due to oxide removal (Fig. 3-15a). Only small roughness variations at the solution/oxide interface were observed by AFM inspection in separate experiments. After 100 s, a transitory reflectance increase can be observed. At t ~ 200 s, reflectance values are reached which correspond almost to those obtained on bare and smooth silicon immersed in an aqueous solution.

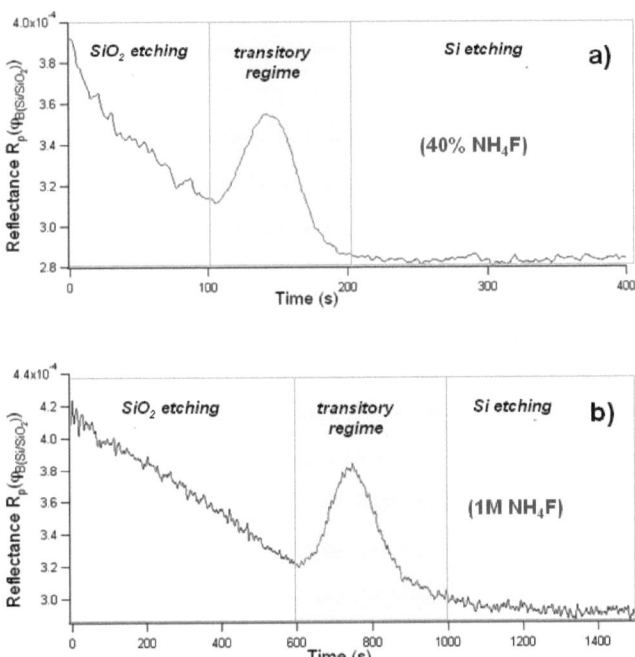

Fig. 3-15: *In situ* BAR monitoring of the SiO$_2$/Si(111) etching behavior in ammonium fluoride solutions: (a) Etching of a native oxide covered Si(111) sample in 40% NH$_4$F. (b) Etching of a native oxide covered Si(111) sample in 1 M NH$_4$F.

In Fig. 3-15b, corresponding results are shown for 1 M concentration of ammonium fluoride. Oxide removal requires 600 s in this case while the transitory reflectance increase extends over a range of 400-500 s. For comparison, H-terminated samples showed only smallest variations of the *in situ* reflectance signal immediately after exposure to the NH$_4$F solution.

Experiments on thermally oxidized FZ-Si(111) were carried out in order to determine a possible influence of the oxide and silicon quality on the observed transitory reflectance behavior (see Fig. 3-16). For these experiments, the samples were adjusted to the bulk Brewster-angle, measured in water. Then, water was exchanged for the etching solution as in the experiment described before. The initial reflectance value (upper curve in Fig. 3-16) corresponds to an oxide thickness of ~18 nm, measured in a solution with an index of refraction of n = 1.33.

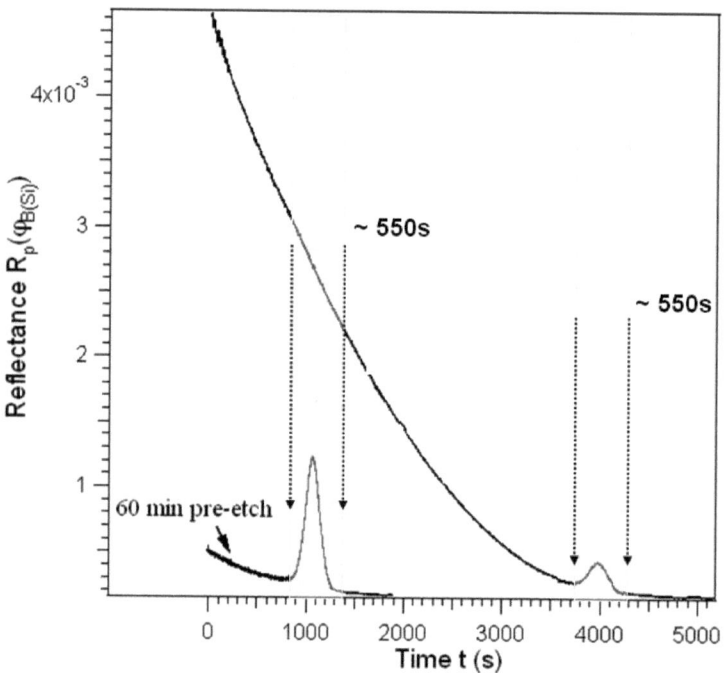

Fig. 3-16: *In situ* BAR monitoring of the SiO_2/ Si(111) etching in 40% NH_4F. Upper curve: SiO_2/Si(111), initially covered with a 18.2 nm thermal oxide layer. Lower curve: SiO_2/Si(111) sample from the same wafer after 60 min pre-etching, measured in renewed solution.

After 3700 s, the reflectance signal is again characterized by a transient behavior for ~ 550 s. After that time, a persisting negative but gradual slope of the reflectance signal behind the transitory region indicates continued smoothening of roughened surface areas. A pre-etch step of 60 min and renewal of the etching solution reduces the oxide etching time as expected until the transitory reflectance regime is observable (lower curve in Fig. 3-16). However, the magnitude of the reflectance in the transitory regime is influenced by the pre-treatment, showing distinctively larger maximum values. The time required for passing the transitory regime (about 550 s) remains unchanged. In repeated experiments, variations of several minutes, necessary for complete oxide removal, were observed. In contrast to these observations, etching times of ultra-thin native oxides exhibited nearly perfect reproducibility. This finding points to concentration gradients and random diffusion processes from and towards the solution/silicon interface that reduces the reproducibility of NH$_4$F-etching of oxides thicker than a few nanometers.

In order to relate the transitory reflectance behavior to the varying density of silicon oxidation states, previously described SRPES results (Figs. 3-6 and 3-7) were reevaluated and shown in Fig. 3-17a and b.

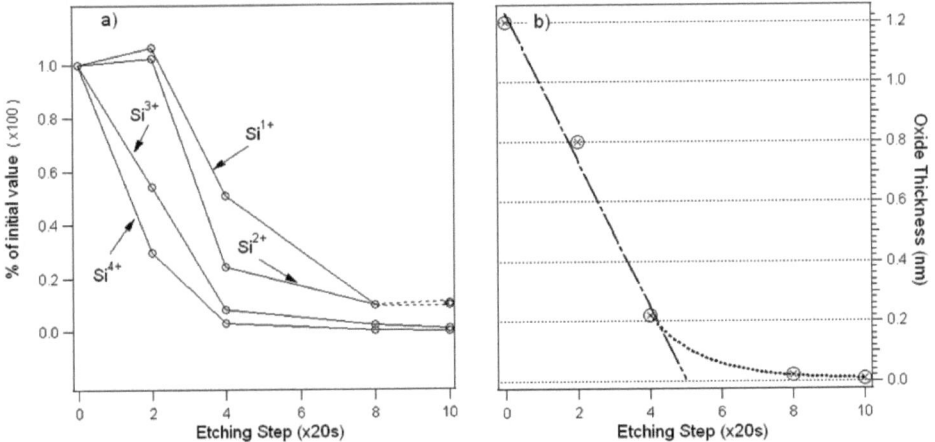

Fig. 3-17: (a) The relative change of Si^{4+}, Si^{3+}, Si^{2+}, and Si^{1+} contributions is shown for the initial 1.2 nm thick native oxide layer and after etching times of 40 s, 80 s, 160 s and 200 s. (b) The change in total oxide thickness during subsequent etching steps is shown. Fitting curves illustrate the oxide etch rate variation (linear until ~ 80 s and exponential for extended etching).

Here, the respective substoichiometric contributions are shown with respect to their initial values before etching. While the contributions of Si$^{3+/4+}$ exhibit a negative slope from beginning on, lower oxidations states are decreasing not until the second etching step. Between 40 s and 80 s, all contributions show a similar relative decrease of their respective magnitudes. Between 80 s and 160 s, lower oxidation states decrease more distinct than Si$^{3+/4+}$. Extending the etching time to 200 s, the Si$^{3+/4+}$ contributions almost vanish while the Si$^{1+/2+}$ signals do not significantly change. Limitations by ambient air sample preparation most likely influence the accuracy of the Si$^{1+/2+}$ assessment for extended etching times. In Fig. 3-

17b, the effective SiO$_2$ thickness, comprising all substoichiometric contributions, is shown. Between etching steps 0 and 4 (corresponding to net etching times 0 s and 80 s) a linear fit was applied and extrapolated to etching step 5 (dashed line). Additionally, an exponential fit was calculated for measurements between 80 s and 200 s (dotted line). Analysis of the O 1s signal indicated small contributions from oxygen in OH for all etch steps. Furthermore, a low F 1s signal was observed for etching times > 4 x 20 s, characterized by a chemical shift due to the presence of Si-F$_x$ species [167].

According to the analysis, it can be assumed that the first two etching steps in 40% NH$_4$F preferably remove silicon oxides in the Si^{3+} and Si^{4+} state. Then, approaching the interface, also silicon in lower oxidation states is etched while the slope of the Si$^{3+/4+}$-curve becomes less steep. Assuming a linear relation between time and oxide etching, a complete oxide removal can therefore be expected after ~100 s of repeated etching. This assumption is illustrated in Fig. 3-17b by a linear fitting curve (dashed line) which suggests an oxide etch rate of 0.6 nm/s. However, SRPES data proves the continued presence of small SiO$_2$ amounts beyond 100 s. Moreover, the assumption of a linear dependence of oxide removal with time is no longer applicable for extended etching. The exponential fitting curve (dotted line in Fig. 3-17b) suggests a slow exponential decrease of the oxide thickness for times > 100 s. The total thickness at 100 s, according to either of the interpolating curves in Fig. 3-17b, is below 2 Å and has to be attributed to interfacial suboxides (Si$^{1+/2+}$) in a sub-monolayer range as well as low amounts of Si$^{3+/4+}$.

While the chemical state of the Si(111) surface during continued etching is assessed by the preceding results, the topographical state remains unclear. Increasing reflectance values in the transitory regime can be attributed either to surface films with increasing thickness or increasing surface roughness. In order to analyze the topographical surface condition the samples were investigated by AFM for selected etching times. First indications for the importance of surface roughness were found in section 3.1.1 as shown in the topographical images of Figs. 3-4a and b. For experiments of continued etching, a combination of measurement techniques was applied: beside *in situ* optical monitoring and AFM surface analysis, dark current measurements were carried out in order to assess the structure of the oxide during etching. For the latter, initial measurements of the open circuit potential (OCP) showed that small anodic currents can be expected during etching if the potential was adjusted to -0.6 V (compared to the OCP of about -1 V for H-terminated silicon). The dark current is related to the reaction of Si with water and characterized by the intermediate formation of Si-OH bonds and subsequent substitution of OH for F [57]. In order to avoid even smallest hole-assisted electrochemical reactions, optical measurements and electrochemical experiments were performed separately. Fig. 3-18 shows two AFM images obtained after emersion at 150 s (near the reflectance maximum) and after 300 s. Surface conditions and corresponding optical and electrochemical behavior are referred to each other by roman numerals, placed into circles. At the reflectance maximum, the surface exhibits aggregations of reaction products, hardly dissolvable by rinsing in UPW. Extended and repeated rinsing was observed to reduce number and size of the structures.

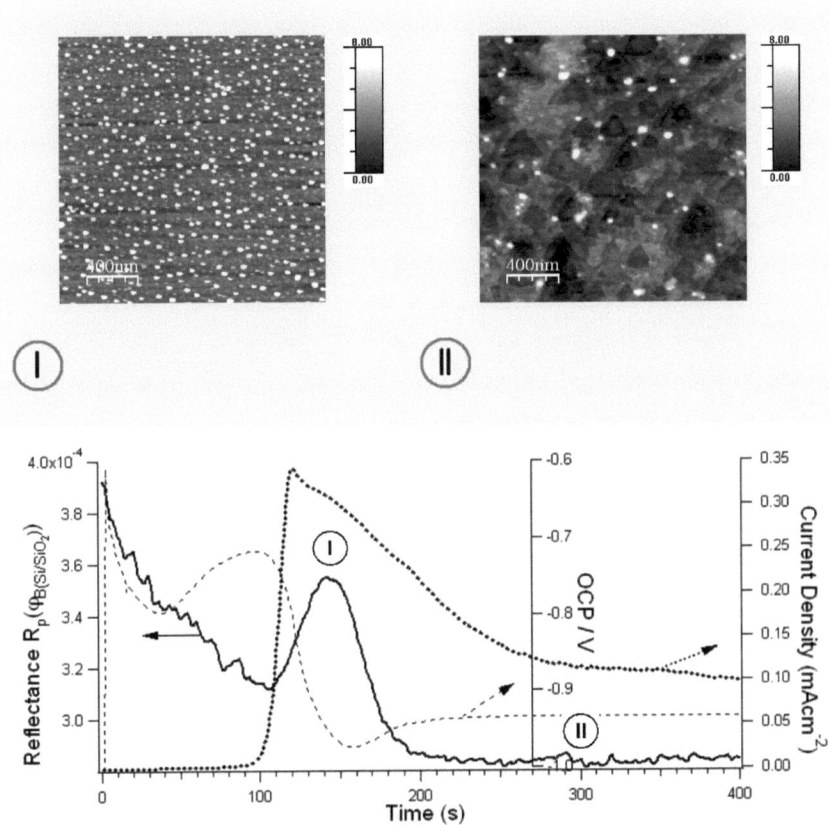

Fig. 3-18: Etching of native oxide covered Si(111) in 40% NH_4F: comparison of *in situ* BAR reflectance with the change of the open circuit potential (dashed line) and the dark current (dotted line), measured at U = -0.6 V. All measurements were carried out in separate experiments. Upper half of the figure: TM-AFM images are shown, corresponding to the reflectance maximum after 150 s (I) and to the surface condition after 300 s (II).

AFM profile analysis showed that the clusters reached several nm in height. Continued etching for 300 s (without intermediate water rinse) initiated the formation of etch triangles, as depicted on the right AFM image.

The structures, seen in Fig. 3-18, left, are not observable before 100 s and are assumed to be consisting of hexafluorosilicate salt $(NH_4)_2SiF_6$ and/or hexafluoride anions SiF_6^{2-}. The formation of theses compounds on NH_4F treated Si(111) surfaces was concluded from photoemission and infrared spectroscopy experiments [168]. Since the solubility of these reaction products in neutral solutions is low, aggregation at the surface may occur if the compounds are built during an elevated silicon dissolution rate. Moreover, the localized distribution on the surface, found after 150 s (Fig. 3-14, left), is assumed to further promote surface roughening by impeding the process of uniform and smooth etching. Incrementing surface roughness upon subsequent etching steps after oxide removal was already observed in the course of *ex situ* Brewster-angle analysis during NH_4F etching of Si(111) (see 3.1.2). In order to correlate surface topography analysis with chemical surface properties, dark current measurements, at a fixed potential, were performed as separate experiment and independently from BAR measurements. As seen in Fig. 3-18, the OCP potential curve starts at $U_{OCP} = -0.6$ V after immersion of the sample into the solution. The following pronounced variation is assumed to result from local *pH* variations during SiO_2 dissolution due to H^+ release, according to the known reaction scheme:

$$SiO_2 + 6\ HF \rightarrow SiF_6^{2-} + 2\ H_2O + 2\ H^+. \qquad (3\text{-}4)$$

Choosing -0.6 V for dark current measurements, the commencement of anodic currents can be interpreted as the beginning process of electron injection into the silicon conduction band on oxide-free surface areas [57]. In Fig. 3-18, results for the optical measurements, dark current and OCP behavior after immersion are compared. The dark current is almost suppressed by the insulating SiO_2 layer until 100 s. The reflectance monitors the decrease in oxide thickness during that time. Strong variations of all three curves are observable between 100 s and 200 s. After 200 s, the curves approach asymptotically their respective steady-state values. For longer exposure times (not shown here), current spikes indicate the abrupt release of hydrogen bubbles from the surface which also cause perturbation of the optical reflectance signal in corresponding *in situ* BAR experiments.

For diluted acidic NH_4F solutions (Fig. 3-19), increasing dark currents were already related to sub-monolayer oxide coverage by photoelectron analysis of the Si 2p core level [169]. For concentrated NH_4F (Fig. 3-18), the almost steep current increase suggests as well increasing oxide-free areas for t > 100 s. At this time reflectance values start to exhibit the transient behavior.

It can therefore be assumed that the transitory reflectance is related to the unimpeded contact of the ammonium fluoride solution with the silicon interface in the course of the SiO$_2$/Si interface dissolution. BAR data as obtained during chemical treatment of the Si(111) samples, i.e. without applied potential, evidence a change of the silicon surface topography after ~100 s (Figs. 3-15 and 3-16) as confirmed by the AFM-analyses in Fig. 3-18. The intermediate increase of the reflectance suggests, following Fresnel's formulae, the formation of an additional layer (adlayer) on top of the surface, causing multiple reflections at the respective interfaces towards the solution and the silicon bulk.

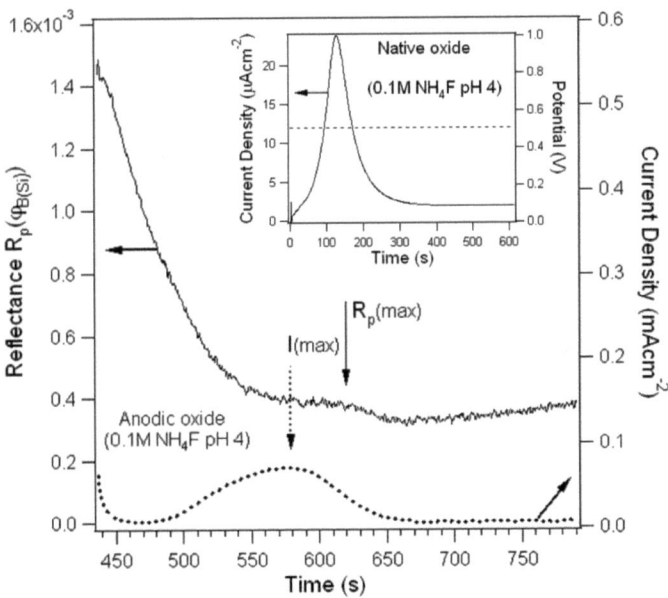

Fig. 3-19: SiO$_2$/ Si(111) etching behavior in acidic ammonium fluoride solution (0.1 M, *p*H 4): simultaneous *in situ* BAR monitoring (solid line) and dark current measurements (dotted line) of anodic oxide etching. The oxide was obtained prior to the presented results at a constant potential of 6 V under white light illumination. Inset: dark current behavior of native oxide covered Si(111) after immersion in 0.1 M NH$_4$F, *p*H 4, at V = 0.5 V.

The composition of this layer may, in principle, comprise surface roughness (considered as an *effective* layer) and/or agglomeration of reaction products. In either case the BAR signal is expected to increase in dependence on layer thickness and on optical properties deviating from those of the underlying silicon. Chemically, the surface region at times around the transitory reflectance regime is characterized by low amounts of sub-oxides in the surface-near region (see Fig. 3-17). Remnant surface oxides as well as buried oxygen precipitates may be the cause of the corresponding SRPES signals. Thermodynamic stabilization of oxygen incorporation into the silicon bulk, as concluded from density functional theory calculations concerning the $SiO_2/Si(100/111)$ interface [5], were discussed before. Oxygen incorporation would increase stress and strain energies stored in the interface region and therefore influence the etching behavior. This consideration necessitates therefore inspection not only of the horizontal effects during etching in the period of the transitory reflectance behavior (see Fig. 3-18) but also of vertical effects, i.e. etch rates.

An etch edge experiment was therefore carried out as illustrated in Fig. 3-20. The surface oxide of two samples was etched in a first step by exposure to 40% NH_4F for exactly 100 s.

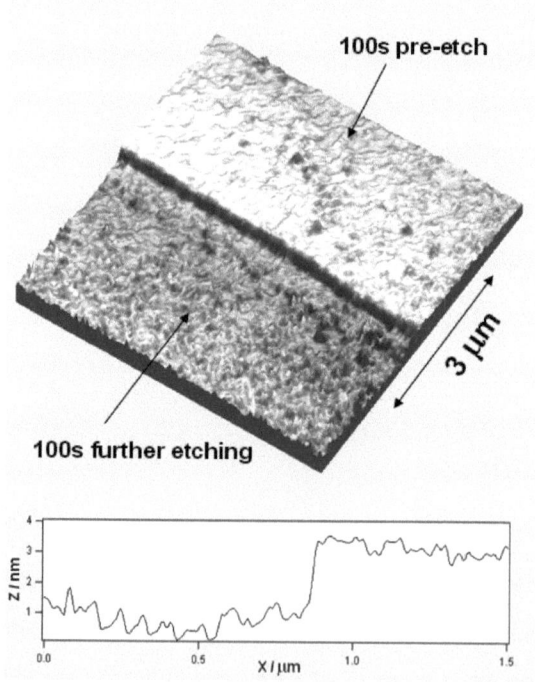

Fig. 3-20: Etch edge of Si(111), after treatment in 40% NH_4F. The surface oxide was removed by a 100 s pre-etch step. The following exposure time of one half of the sample to the solution was 100 s. Height analysis (lower picture) reveals a step height of ~ 3 nm.

Subsequently, the samples were rinsed in UPW and one half of one silicon sample was covered by the second sample. Beside increased wetability, pronounced adhesion of the sample faces was observed. The position of the samples was fixed by a pair of stiff tweezers. After further etching for 100 s, the samples were detached and the resulting etch profile was measured by AFM as shown in Fig. 3-20. Comparable increased wetability was reported recently for partially etched silicon oxides after treatment in diluted hydrofluoric acid [170]. The finding was attributed to fluorinated silicon-oxygen compounds that adsorb water effectively and leads to increased bond energies for bonded pairs of silicon-oxide-covered wafers. In the case described here probably lower valence states ($Si^{1+/2+}$) contribute to the adhesion in a similar manner.

Fig. 3-20 shows that a ~3 nm deep Si layer is etched between t =100 s and 200 s, corresponding to an enhanced (average) etch rate of 1.8 nm/min, while the SRPES SiO_2 signal slowly decays from ~ 2 Å to below 0.2 Å. Hence, the integrally measured SiO_2 signal behaves almost linear as long as the interface is not reached (until ~80-100 s). Then, remaining (and possibly bulk incorporated) oxides in the 3+/4+ valence state are removed subsequently at a lower rate in the course of beginning silicon dissolution. Simultaneously, remaining interfacial suboxides ($Si^{1+/2+}$) are etched further, leaving behind increasing oxide-free areas on which enhanced electron injection is observable in separate dark current measurements (Fig. 3-18).

For initially SiO_2 covered Si(111), etch rates in concentrated ammonium fluoride have been published suggesting a low etch rate of 0.45 nm/min in the limit of extended (bulk) etching [163]. Compared to this rate, the measured value (~1.8 nm/min) is about four times higher and suggests accelerated silicon etching after the top-surface oxide has been removed. The increased etch rate has to be attributed to stress and strain energies stored in the interface-near region. Consequently, reaction products agglomerate in larger quantities on the surface during this period of time.

Detailed modeling of the etching process, comprising oxide etching, the transitory reflectance regime and bulk etching can finally be achieved by evaluation of the optical data in terms of multi-layer analysis. Pure etching of the stressed layer, aggregation of reaction products and surface roughening effects can thereby be distinguished. Optical properties and respective thicknesses of the oxide layer, the stressed / strained Si layer and their respective interfaces were taken into account. Optical data, describing the properties of bulk silicon at 500 nm wavelength were again chosen from the literature [95]. Subsequent etching of 1.2 nm SiO_2 with a roughness of rms = 0.3 nm at the solution/SiO_2 interface was assumed, followed by 3 nm Si-interlayer etching with comparable roughness at the interface towards the SiO_2 layer as well as slightly increasing roughness at the interface to bulk silicon. On approaching the respective interfaces, sub-monolayer coverage was modeled by Bruggeman effective medium theory. The dielectric function of the interlayer was chosen according to *ex situ* analysis in section 3.1.2. The thickness of the interlayer, experimentally determined by AFM measurements is 1.5 nm smaller than obtained by multi-layer analysis evaluating *ex situ* BAA data of subsequently etched silicon. Using these data, the formation and removal of the topmost surface coverage during interlayer etching is evaluated in Fig. 3-21: the simulated reflectance signal at the observation angle $\varphi_{B(Si/SiO2)}$ is presented as dotted line in Fig. 3-21a and compared to the measured reflectance signal of the SiO_2/Si system immersed in 1 M NH_4F (see Fig. 3-15b).

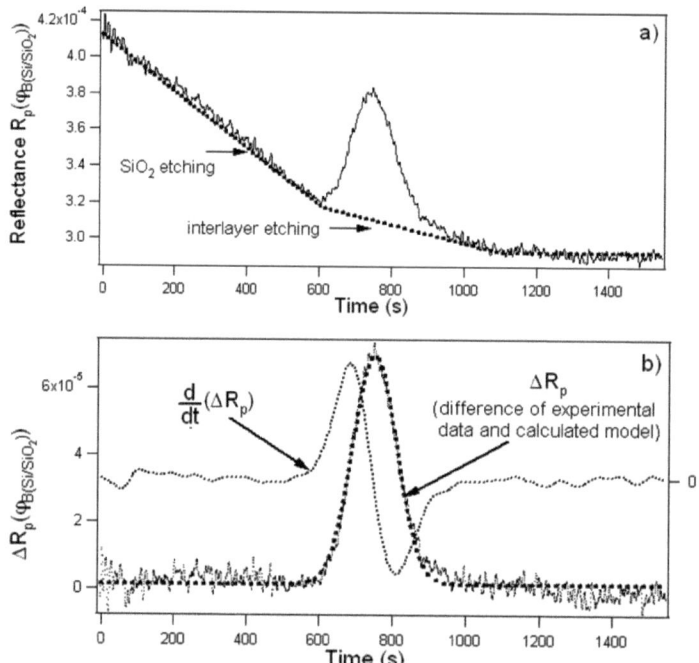

Fig. 3-21: Multi-layer analysis of oxide and interlayer etching in 1 M NH_4F. (a) Experimental data (solid line) compared to simulated data (dotted line) according to a multi-layer model comprising 1.2 nm surface oxide, a 3 nm Si interlayer and additional interface roughness layers. No aggregation of precipitates or increasing roughness in the regime between 600 s and 1200 s was assumed. (b) Difference of the experimentally determined curve and the calculated reflectance. The reflectance difference curve, ΔR_p, was fitted by a Gauss function (dotted line). Additionally, the time derivative of ΔR_p is shown.

The respective reflectance regions behave almost linear, providing thus a background-like curve for pure surface effects between ~600 s and ~1100 s. After subtraction, a signal, ΔR_p, is obtained which resembles a Gauss distribution curve, representing the formation and removal of both surface roughness and aggregation of reaction products (Fig. 3-21b).

Eventually, this type of function was used to carry out a least-square approximation shown as dotted line and compared to the subtracted signal. Good agreement between experimental data and approximation is achieved. However, a slightly increased error is obtained around t = ~900 s and t > 1200 s, suggesting that the reflectance R_p is rather characterized by a slowly decreasing exponential tail after passing the transitory region (see Fig. 3-21a) which was not considered by the approximation. It can be therefore assumed that the interlayer extends deeper into the bulk and suggests that the interlayer does not completely dissolve (with decreasing etch rate) before t_E ~ 10 min in agreement with *ex situ* BAA.

The numerical simulation elucidates, too, the dependence of the optical response to the (electro-)chemical process: in the limit of thin films, the reflectance $R_p(\varphi_{B(SiO_2/Si)})$ behaves almost proportional to the layer thickness. The time derivative of R_p and ΔR_p in the transitory region is therefore related to the rate at which agglomeration/roughening occurs (dotted line in Fig. 9b). In electrochemical experiments, this rate is related, in turn, to the dark current, accounting thus for the phase shift of reflectance and current (Figs. 3-15 and 3-18) because the former behaves like the time integral of the latter (see analysis in 3.2.3). Negative values of $d/dt(\Delta R_p)$ consequently indicate the rate at which reaction products are removed. Due to the symmetry of the derivative, formation and dissolution of the adlayer (the agglomerates) appear to be the result of the same reaction mechanism, i.e. silicon etching.

Initial pitting may therefore be induced by (under-)etching of the agglomerates, leading finally to etch triangles as depicted in Fig. 3-18, right, and Fig. 3-23, left. Suppression of pit formation was observed by intermediate UPW rinsing (Fig. 3-23, right), carried out after 100 s etching of the top-surface oxide layer. The aggregation probability of reaction products decreases by this treatment and has to be assessed in relation to the increased polarity of the surface, after the bulk oxide was removed, and the reinstated near-surface *p*H value by use of a fresh NH_4F solution.

A schematic evaluation of the interlayer etching process, carried out for thermally oxidized silicon, is finally presented in Fig. 3-22. A magnification of the BAR data (Fig. 3-22a) is shown together with corresponding surface/interface conditions of the multi-layer silicon system evolving with increasing etching time. The initial system comprises again the surface oxide, a silicon interlayer beneath the SiO$_2$/Si interface, the silicon bulk and the respective interfaces (Fig. 3-22b). After SiO$_2$ removal, the transitory reflectance behavior corresponds to a decreasing interlayer thickness and formation of surface roughness as well as aggregation of reaction products (Fig. 3-22c). The extended time of adlayer formation (5 times larger than observed for native oxide) can be understood by the slower etch rate of the denser oxide. Consequently, the average interlayer etch rate decreases due to the slower exposure of the silicon interface to the solution. Upon reaching bulk silicon, a small negative slope of the reflectance indicates the presence of surface roughness and remnants of the interlayer which are gradually reduced by subsequent etching (Fig. 3-22d).

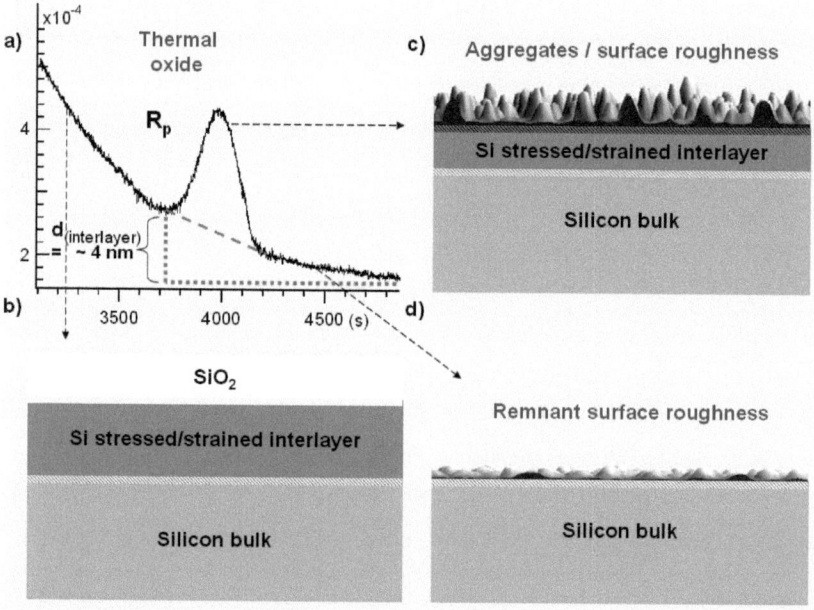

Fig. 3-22: Schematic representation of the SiO$_2$/Si and Si-interlayer dissolution process according to the model considerations described in the text. (a) *In situ* BAR data of a thermal oxide covered Si(111) sample; arrows indicate the relation to the respective multi-layer schemes. (b) Model of the SiO$_2$/Si-interlayer/Si system during SiO$_2$ etching. (c) Model of the partially etched and roughened Si-interlayer with aggregation of reaction products. (d) Model of the Si bulk with surface roughness and remnants of the stressed interlayer.

According to the model considerations given above, the interlayer thickness is approximately related to the reflectance difference between the offset of the transitory period and a steady-state value obtained for extended etching. Applying multi-layer analysis, an approximate thickness of 4 nm for thermally oxidized silicon was calculated. This value is in good agreement with *ex situ* BAA analysis and confirms that extended etching has to be employed in order to remove the stressed interlayer and to obtain high-quality Si(111) surfaces.

3.1.4 Morphological and chemical optimization of Si(111)-1x1:H

In the preceding section, the importance of chemical and topographical transitions during etching of Si(111) in concentrated NH_4F was shown. For the regular surface topography, shown in Fig. 3-10, no deoxygenation of the etching solution was necessary. Only an intermediate UPW rinse seemed to suppress the formation of triangular etch pits. However, the chemical state of the H-terminated surface (beyond the transitory reflectance regime) was not assessed with highest surface sensitivity (SRPES). Since the role of dissolved oxygen for the resulting Si(111) surface morphology found wide agreement in the literature [139], the previous findings were reproduced. In Fig. 3-23, the two surface topographies after 10 min etching (with and without intermediate water rinse) are compared as observed by CM-AFM.

Fig. 3-23: CM-AFM images of Si(111) surfaces, initially covered with native oxide. Left: after 10 min continued etching. Right: after 100 s pre-etching, subsequent UPW rinsing, N_2 drying and continued etching for further 10 min.

The comparison of topography-formation and *in situ* optics shows only one but probably important difference for both treatments. As depicted in Fig. 3-24, the intermediate cleaning of the surface at a time where top-surface oxides are almost completely removed results in a faster passage through the transitory region. The reflectance measurements at a constant angle of incidence (the Brewster-angle $\varphi_B(SiO_2\text{-}Si)$ is not sensitive enough in order to distinguish between the two different forms of roughness on the respective surfaces as shown as insets. Intermediate UPW rinse removes reaction products from the interface, exposed by SiO_2/Si interface dissolution, and facilitates subsequently uniform silicon etching in the renewed NH_4F solution with reinstated concentration and pH. This result follows from the analysis of sections 3.1.2 and 3.1.3.

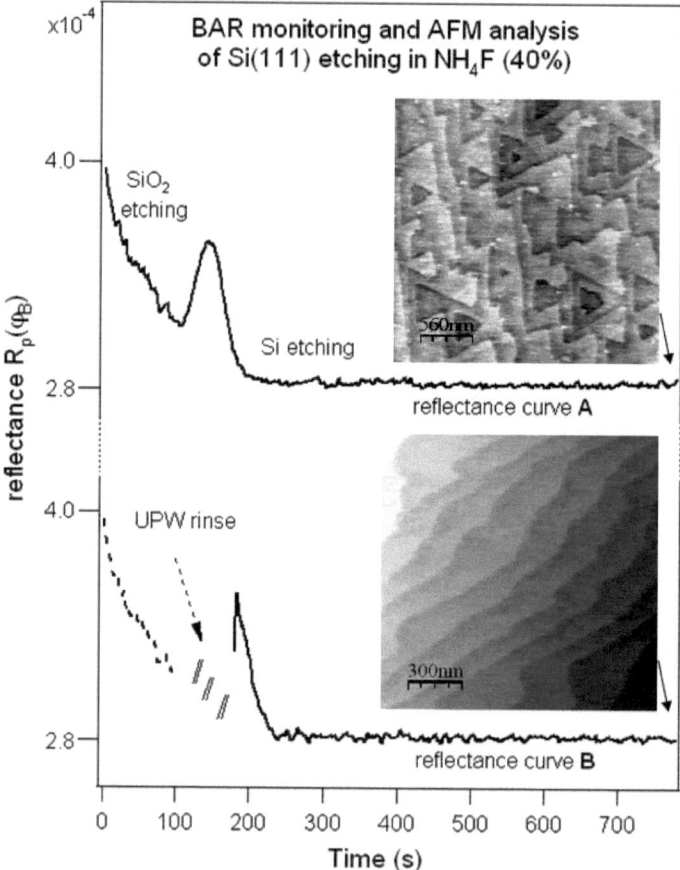

Fig. 3-24: Etching behavior of Si(111), initially covered with native oxide, in NH_4F (40%) as monitored by *in situ* reflectance measurements (BAR). Topographical analysis was performed by AFM at the times indicated by arrows. Reflectance curve A was recorded for continued etching for 800 s; reflectance curve B was measured after 100 s pre-etching and following UPW rinse. The dashed part in curve B between t = 0 and 100 s is the same as in curve A.

The chemical state of the surface, prepared with intermediate UPW rinsing, was analyzed by SRPES with an excitation energy of hv = 150 eV. In Fig. 3-25, left, the measured Si 2p envelope is shown, comprising the bulk Si signal and varying oxidation states between 100 and 103.1 eV. After Shirley background subtraction and Lorentz-Gauss peak profile characterization, the respective contributions to the envelope curve were calculated by a deconvolution procedure and indicated as percentage of the total integral of the curve. Almost no SiO_2, corresponding to Si^{4+}, is detectable while low amounts of Si^{1+}, Si^{2+} and Si^{3+} were found corresponding to Si_xO_y in varying stoichiometric composition. A low F 1s detection signal (not shown here), measured at 750 eV excitation energy, suggests that also compounds as =Si-H-F and -Si-H-OH-F have to be considered with two and one Si back-bonds, respectively. These contributions result in chemical shifts of about 1 and 1.5 eV, according to earlier DFT calculations [171], and interfere with Si_xO_y in the Si 2p core signal which impedes accurate quantitative analysis. Comparable results were obtained for a sample analyzed immediately after the transient behavior of the BAR reflectance signal (reflectance curve A at t = 200 s in Fig. 3-24), i.e., in the beginning of Si bulk etching.

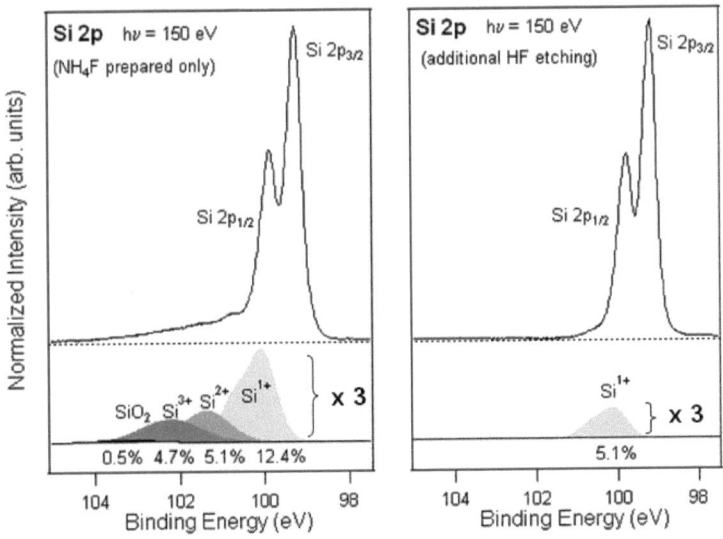

Fig. 3-25: SRPES analysis of the Si 2p core level. SiO_2 and substoichiometric oxide contributions were magnified for clarity by a factor of three. Left: Si(111), etched in NH_4F (40%) for 100 s and 10 min with intermediate UPW rinse and renewal of the etching solution. Right: Si(111) after an additional HF (50%) etch step for 10 s and final UPW rinse. Percentages refer to the total integral of the envelope curve. $Si^{2+/3+/4+}$ contributions were too low to obtain reliable values.

Therefore, the presence of minute Si^{x+} amounts is considered to be independent of the etching time and to result from the inherent =Si-H-OH(F) formation during silicon etching in NH_4F [57, 139]: according to this scheme, a Si atom at atomic step edges with two H-saturated dangling bonds reacts with water to form =Si-H-OH as an intermediate. Subsequently, OH is substituted for F. A partially OH/F-terminated surface is thereby left behind after emersion from the solution. Polar bonds such as Si-OH / Si-F reduce the hydrophobicity of the surface and possibly promote further oxidation during UPW rinse [172] and exposure to the ambient.

In a separate experiment, the NH_4F prepared surface (after rinsing in UPW and N_2 drying) was additionally etched in 50% HF for 10s. The corresponding SRPES analysis in Fig. 3-25, right, shows that contributions of lower oxidation states are distinctively smaller; particularly, the amounts of Si^{2+} through Si^{4+} fell below 0.1% and are not indicated in the figure due to low reliability. The interaction with HF therefore increases the H-termination at atomic steps. Furthermore, the surface may be less susceptible to unwanted oxidation during the short periods of subsequent UPW rinsing and intermediate contact to the ambient air. Concentrated HF is reported to leave less fluoride species on the surface than diluted HF solutions [172]. Correspondingly, only a low F 1s signal, shown in Fig. 3-26, could be measured. The deconvolution shows contributions from F as well as Si_xF_y.

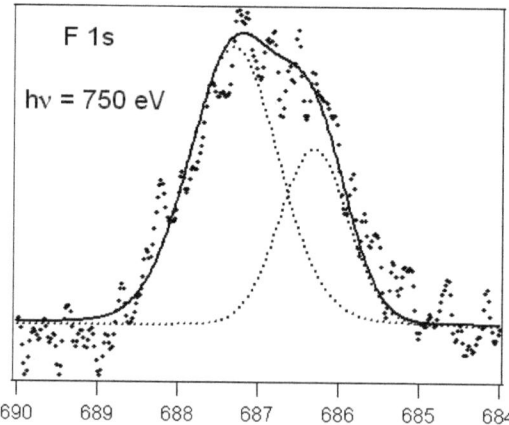

Fig. 3-26: F1 s signal measured on a Si(111) sample after additional treatment with 50% HF for 10 s.

Since the remnant Si^{1+} signal does not allow a differentiation between oxygen in Si-OH or in Si_2O [173], an analysis of the O 1s core level was carried out before and after HF etching (Fig. 3-27). Due to the increased O 1s core hole screening by electrons of the surrounding Si atoms, oxygen in Si_2O appears at a lower binding energy (near 531.5 eV) than oxygen in the Si-OH configuration (near 532.4 eV) [174]. From Fig. 3-27 follows that oxygen in Si-OH is still detectable as well as oxygen in Si_2O, i.e., oxygen appears as a bridging atom in Si-O-Si. In comparison to the chemical state before HF etching, the respective integrals of the deconvoluted curves show a decrease by a factor of 2.5 for the Si-OH signal and 3.0 for the Si_2O signal.

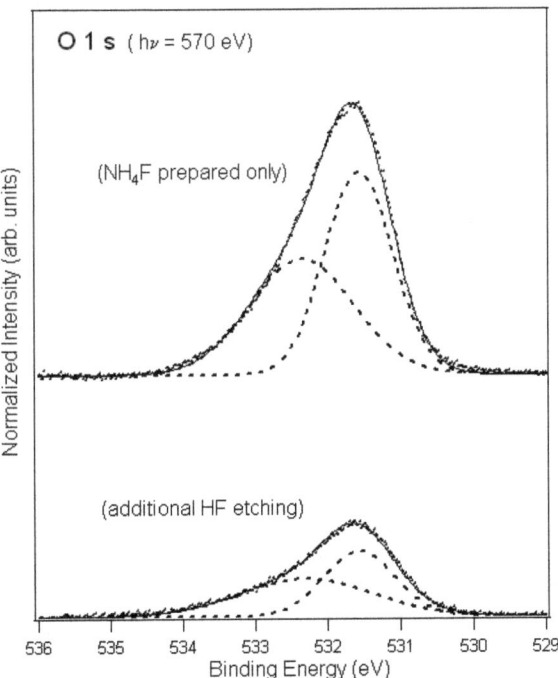

Fig. 3-27: O 1s signal analysis for the two treatments described in Fig. 3-24 and in the text. After HF etching, the integral of the curve has decreased by a factor of ~ 2.7. The peak line at E_B ~ 531.5 eV is mainly attributed to the presence of minute amounts of silicon in the Si-O-Si configuration. At a higher binding energy (E_B ~ 532.4 eV), remnant Si-OH coverage is detectable. Both signals contribute to the Si^{1+} peak line in Fig. 3-25.

Topographical effects of the finishing HF etching step are shown in Fig. 3-28: the two micrographs obtained by contact-mode AFM compare the stepped topographies before (left) and after HF etching (right) on the same sample. At the scale of these images (5 μm x 5 μm), no surface roughening (rms roughness ~ 0.4 nm) or step disorder after the HF treatment is observable although the rather large Si^{x+} signal in Fig. 3-25, left, suggests oxidation also on flat terraces. The less pronounced parallel alignment of the terraces is attributed to inhomogeneities of the native oxide. The miscut angle, mainly oriented towards the $<11\bar{2}>$ direction would therefore slightly vary and influence the anisotropic etching behavior after oxide removal. Such *aging* effect, i.e. continuing native oxide growth, was observed on samples stored in the laboratory for about ten years.

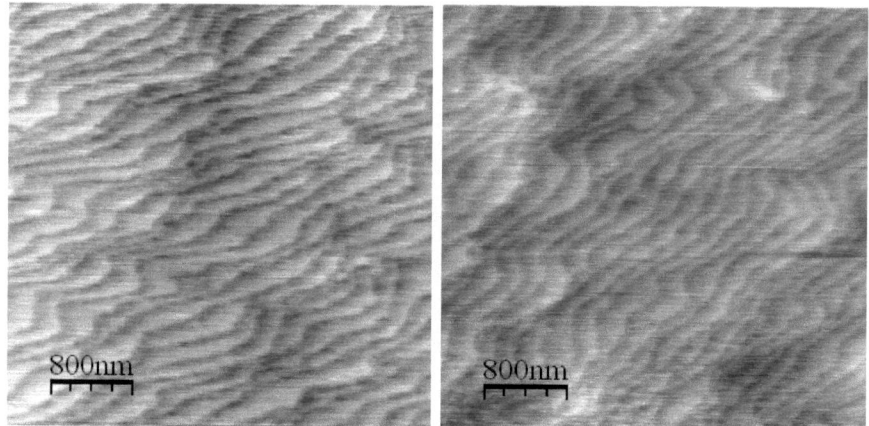

Fig. 3-28: Contact-mode AFM images before (left) and after (right) the final HF (50%) etch step, measured on the same sample. The images were recorded at different but representative sites of the Si(111) surface which is characterized by a miscut angle towards $(11\bar{2})$ direction. Almost no pitting effects are observable. The root mean square roughness of ~ 0.4 nm is not affected by the final HF treatment.

Further evidence for the presence of oxidized silicon after etching in NH_4F was sought by UPS measurements of the cut-off of secondary electron emission, using a bias voltage of 5 V. In Fig. 3-29, three surface conditions are compared. FZ-silicon samples were used in order to minimize the oxygen content caused by the silicon growth technique:

1) Thermally oxidized FZ-Si(111) (dashed curve) was removed from the solution at a time where the transitory reflectance behavior is close to its maximum (see Fig. 3-16). The time was chosen in order to determine the chemical surface condition that causes, on one hand, the observed increased wetability and, on the other hand, the composition of reaction products that form nanometer-seized structures (see Fig. 3-18). A pre-etching step of 60 min was

applied to produce a condition that would correspond to maximized reflectance as shown in Fig. 3-16.

2) H-terminated FZ-Si(111), exposed to NH$_4$F only, was prepared by 100 s pre-etching and additional 10 min etching (dotted curve).

3) The third FZ-sample was additionally exposed to concentrated HF (50%) for 10 s to produce the optimized surface condition discussed above.

The cut-off of secondary photoelectrons by He I (21.8 eV) and He II (40.2 eV) excitation showed significant differences for the three samples. At the cut-off, the condition for the kinetic energy, E_{kin}, is given by $E_{kin} = 0$ eV. Steps in the resulting spectrum indicate that the electron affinity is lowered on some areas. This effect is very pronounced for samples 1 and 2. The observed energy shift is larger than 1 eV for both samples. The step height is larger for sample 1, originally covered with thermal oxide. TM-AFM analysis of the surface was impeded by a film that blurred the image indicating strong tip-sample interaction on large surface areas. It appears therefore likely that the step height can be conceived as indicator for the fraction of the surface area that is covered by the dipole layer.

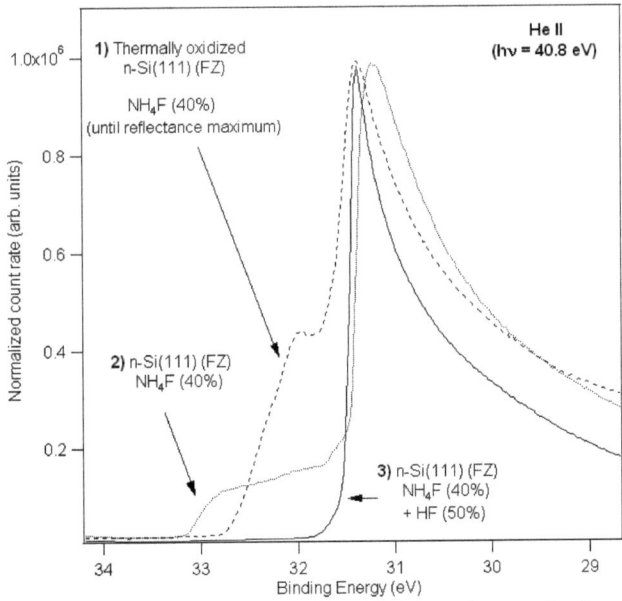

Fig. 3-29: Measurement of the cut-off energy of secondary electrons for three different surface conditions. (1) Thermally oxidized FZ-Si(111) after removal of the oxide layer. (2) H-terminated Si(111) prepared in NH$_4$F (40%) with stepped surface topography. (3) Sample preparation as in (2) with additional dip in 50% HF. A bias voltage of 5 V was applied.

The corresponding Si 2p signal for this sample, measured with Mg K$_\alpha$ excitation (1253.6 eV) indicates no silicon dioxide while the O1 s signal is detected with increased count rate (see Fig. 3-30).

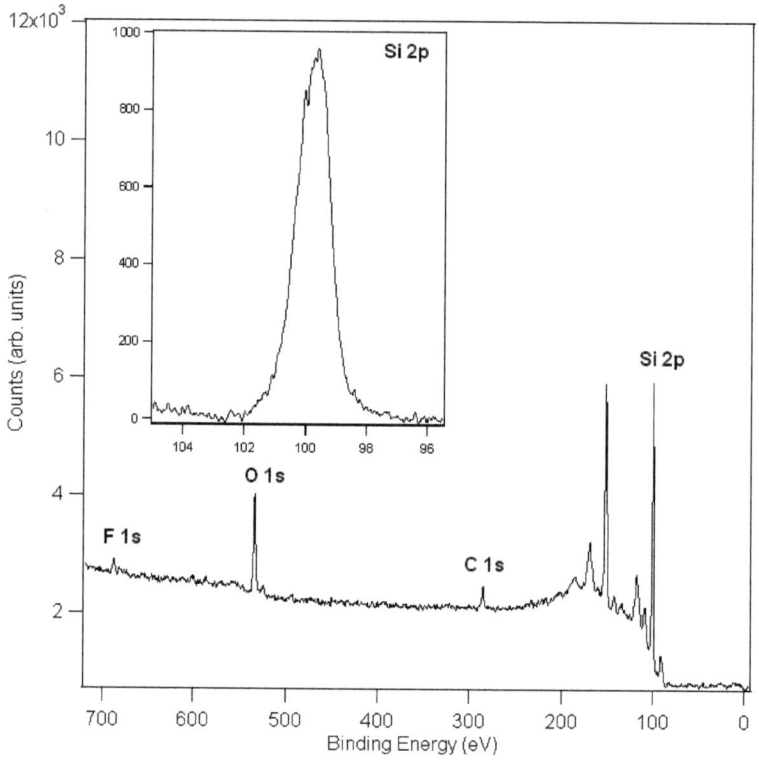

Fig. 3-30: Mg K$_\alpha$ XPS analysis (1253.6 eV) of the FZ-Si(111) sample after removal of the thermal oxide (18.2 nm) and before etching of the stressed layer. The survey spectrum is shown for energies between 0 and about 700 eV. The inset shows the Si 2p core-level signal.

For sample 3, additionally etched in HF, no step in the cut-off region is visible. The O 1 s is approximately as large as for the only NH$_4$F prepared surface. After background subtraction, the O 1s to Si 2p ratio is 0.3 for samples 2 and 3 but 0.6 for sample 3. The analysis of the O 1s signal for all surface conditions suggests that oxygen is present in the Si-OH configuration rather than in Si-O-Si bridging molecules. The polar character of the Si-OH compound is interpreted as the cause for the dipole layer that lowers the electron affinity at some sites on the surfaces. The intermediate exposure of the sample to the ambient (which was inevitable in these experiments) may have increased the dipole formation by adsorption of water molecules from air with subsequent orientation of the water dipoles in the presence of Si-OH bonds. This effect is least significant for sample 3 since the H-termination is almost complete. In contrast to the assumption that hexafluorosilicates are forming on the surface during the transitory reflectance behavior, no nitrogen, as part of the molecule composition, could be detected neither fluoride in higher amounts. While the increased wetability of the Si(111) surfaces after removal of the top-surface oxide and before etching of the strained layer is tentatively explained by the amount of Si-OH bonds, the chemical nature of agglomerated reaction products, as shown in Fig. 3-18, is not completely clarified. According to these results, the morphological perfection of NH_4F treated Si(111) surfaces benefits from an appropriate pre- and post-etching treatment. Two effects are thereby achieved:

1) The formation of triangular etch pits can be restrained by intermediate UPW rinsing while the H-termination at atomic step edges is improved by a finishing HF (50%) dip.
2) Surface dipole layers, distributed in the sub-monolayer range across the surface, are furthermore removed.

3.1.5 Synopsis: chemical and structural properties of the stressed interfacial region

The properties of the stressed and strained layer, buried beneath oxide overlayers on Si(111) surfaces, were analyzed with respect to their chemical and structural nature. With surface sensitive SRPES analysis, the vertical distribution of silicon in different oxidation states was investigated upon successive etch steps in NH_4F (40%). Horizontal and vertical topographical effects were investigated by AFM surface- and etch-edge-analysis. Integrally measured chemical and structural interface properties, obtained by BAA and BAR, could be disentangled by the combinatorial approach of the used methods. The results suggest that stress forces, exerted by the oxide overlayer and bulk-incorporated oxygen atoms, affect the stability of the silicon substrate over a range of about 3 – 4 nm. The depth of the layer depends thereby on the density of the oxide layer. Consequently, altered optical properties and accelerated dissolution rates of this region were observed. The fact that the formation of triangular etch pits could be related to this accelerated dissolution (by underetching of reaction products) is opposed to analyses in the literature where oxygen, dissolved in the NH_4F solution, is discussed. It could be shown, by application of *in situ* BAR, that the findings are independent of the solution concentration. As a main result, an optimized Si(111) preparation method could be developed and further improved after SRPES and UPS investigations of the chemical surface state and electron affinity characteristics. The observation of surface dipoles, present after NH$_4$F preparation (and not after an additional HF dip), necessitates further investigations in the future with well-defined experimental conditions and theoretical model considerations. For experimental results to be discussed in the following section both findings, stress-induced accelerated substrate dissolution as well as the influence of Si-OH terminated bonds after NH$_4$F etching, will be of importance.

3.2 Horizontal nanostructure formation by photoelectrochemical conditioning

Micro- and mesoporous silicon formation by anodic HF etching generally transforms bulk silicon into a sponge-like structure of interconnected and hydrogen-covered silicon columns [175]. The actual surface and sub-surface geometries strongly depend on doping, etching conditions and, for n-type material, illumination intensity [176-178]. Corresponding to the large variety of experimental findings, numerous pore formation and propagation models are discussed such as point defect supersaturation [179], virtual passive layer formation at the pore walls [180] or the influence of doping atoms [181]. For the nucleation of pores, hydrogen incorporation in a near-surface region was suggested [182]. Chemical reaction schemes, in turn, were proposed by different authors [183-185]. Of particular importance, in connection of the preceding chapter 3.1, are strain effects observed for both freshly prepared porous silicon and aged, i.e. oxidized porous silicon (PS). For freshly prepared PS, the PS lattice parameter is slightly expanded in a direction perpendicular to the surface while the corresponding parameter parallel to the surface is that of bulk silicon [186]. The strain measured perpendicular to the surface is of the order 10-40 x 10^{-4} [187, 188]. A further increase of strain is observable after growth of a thin oxide layer. Wetting strain was observed when PS was filled by a liquid. Strain is attributed to the presence of Si-H$_x$ on the large internal surface of the pores [188] and increases for oxidized samples to 10^{-2}; stress forces increase from 0.01 GPa to 1 GPa. Meanwhile, the application of PS [37] or oxidized PS [189] is discussed as stress generating nanomaterial, less expensive than, for instance, Si$_{1-x}$Ge$_x$ epitaxial layers [190]. Generally, PS is analyzed in terms of sub-surface topographical properties, pore shape and density. At the semiconductor-electrolyte interface, however, structures are observable which protrude out of the surface. Due to the structure size of these nanoparticles, those effects as quantum confinement and Coulomb blockade can be achieved. Silicon based single-electron transistors [191] and quantum-dot floating gate memories [192] are, for instance, attractive applications and intensify corresponding investigation efforts [193, 194]. Typically, the Si(100) surface orientation is investigated, exposed to diluted solutions of HF [193, 194]. Studies of the Si(100)/NH$_4$F are comparably rare [195]. The initial dissolution process of Si(111) in diluted NH$_4$F was addressed in a few publications [196-198].

In combination with *in situ* Brewster-angle reflectometry, the photodecomposition of n-type Si(111) in diluted NH$_4$F containing solutions is analyzed in the following. In the divalent dissolution regime, the observed current-voltage shift upon incremented illumination intensity is exploited by selective oxidation and shaping steps, subsequent to self-organized nanostructure formation. Model considerations tentatively explain the findings by enhanced oxidation of structures at the porous silicon / silicon bulk interface with pointed curvature. The stress induced growth of oxide leads to an observable change in the aspect ratio of the structures. In the oscillatory regime, real-time monitoring of the oxide layer growth is combined with quantitative analysis of oxide thickness as well as SiO$_2$/Si interface roughness. At the interface, micro- and nanostructures are observed which suggest likewise self-organization principles and the influence of oxide induced stress fields. In model considerations, the well-known S-shaped behavior of the dissolution valency, from divalent to tetravalent dissolution, is related to varying dissolution branches which lead to either anisotropic terrace removal and/or pore formation in the course of increased SiO$_2$ formation.

3.2.1 Alignment effects and shaping of nanostructures in the divalent dissolution region

The typical photocurrent-voltage curve of n-Si(111) and the corresponding BAR signal are shown in Fig. 3-32. The current is obtained upon potential increase from -0.6 V to 10 V at a scan rate of 5 mV/s. Between OCP and the first current maximum the overall dissolution reaction is divalent and can be described as:

$$Si + 6HF + h_{VB}^+(h\nu) \rightarrow SiF_6^{2-} + 4H^+ + H_2 + e_{CB}^- . \tag{3-4}$$

In this region, the BAR signal indicates the formation of a layer that may comprise, according to Fig. 2-5, both roughness and porous silicon formation. Optical analysis according to Eq. 2-27 shows the existence of a rough layer of ~ 2 nm thickness. Interestingly, a shift between reflectance and current maximum can be observed: the reflectance maximum precedes the corresponding photocurrent maximum. Photoelectron spectroscopy using synchrotron radiation and current transient measurements proved earlier that oxidation, at a low reaction rate, occurs before the first current maximum [171, 199]. BAR data seem to confirm these results if it is assumed that the rough surface topography is partly substituted, at higher potentials, by the electrochemically grown oxide. Due to the similarity of the corresponding refractive indices of the electrolyte (n ~ 1.33) and the oxide (n ~ 1.38), the surface would appear less rough at the current maximum where the oxidation rate is already increased. However, constant-reflectance experiments, to be introduced below, show that the interplay of top-surface and sub-surface topographic features, i.e. surface roughness and porous layer formation, proceeds probably more intricate and a dynamic interaction of the dissolution branches (divalent and tetravalent) has to be considered before the first current maximum.

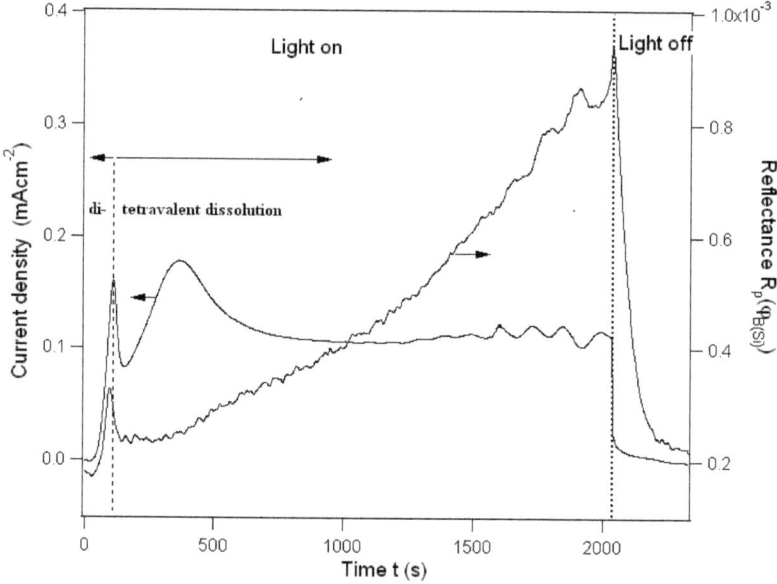

Fig. 3-32: Photocurrent-voltage characteristics for n-Si(111) in 0.1 M NH$_4$F, pH 4, scan rate 5 mV/s, and *in situ* reflectance data measured at φ$_B$(Si). After 10 V were reached (~2000 s), measurements were continued in the dark.

At higher anodic potentials, the current starts to oscillate due to the competition between oxide formation and etching which results in a dynamic equilibrium situation. Here, the reaction is tetravalent and can be described as:

$$Si \; + \; 6HF \; + \; 4h^+_{VB}(h\nu) \; \rightarrow \; SiF_6^{2-} \; + \; 6H^+ . \tag{3-5}$$

The reaction described by Eq. 3-5 comprises, firstly, an oxidation step and, secondly, a chemical etching step which depends on the *p*H value of the solution:

$$Si + 2H_2O + 4h_{VB}^+(hv) \rightarrow SiO_2 + 4H^+ \quad \text{(oxidation)}, \quad (3\text{-}6)$$

$$SiO_2 + 6HF \rightarrow SiF_6^{2-} + 2H_2O + 2H^+ \quad \text{(chemical etching by HF)}, \quad (3\text{-}7a)$$

and $\quad SiO_2 + 3HF_2^- \rightarrow SiF_6^{2-} + H_2O + OH^- \quad$ (chemical etching by HF_2^-). (3-7b)

Eqs. 3-7a and b are describing the process of chemical etching of SiO_2 in dependence on the *p*H value of the NH_4F solution which determines the dissociation of NH_4F into HF, HF^{2-} or F^- [200]. Simultaneously, the oxide etch rate is affected, too, since HF_2^- etches much faster than HF. The oscillation cycles are again characterized by a phase shift of the reflectance with respect to the photocurrent. In this case, however, photocurrent maxima precede those of the reflectance. At V = 10 V, the W/I$_2$ lamp was switched off and measurements were continued in the dark at V = 0.5 V. The dark current drops immediately and shows in the following the typical transient behavior of oxidized silicon samples, i.e., chemical etching removes the oxide layer, and interfacial suboxides are subsequently dissolved in an electrochemical reaction. Fig. 3-19 shows a magnification of the dark current, and the transient behavior can clearly be observed at this magnified scale. The rapid decrease of $R_p(\varphi_B)$ in Fig. 3-32 corresponds, in turn, to the fast etch rate of the anodic oxide in 0.1 M NH_4F (*p*H 4).

Figure 3-33, left, shows the behavior of the photocurrent for different light intensities. It is remarkable that the OCP shifts by ~ 240 mV towards more cathodic values when the intensity is increased from 1 to 40 mW/cm^2. As a consequence, the electrochemical reaction can be manipulated by variation of the light intensity: starting with a low light intensity (~ 1 mW/cm^2) at a potential of -0.1 V, photocurrents are expected to commence at position 'A' in Figure 3-33, left. The reaction proceeds by divalent dissolution at the increasing part of the photocurrent curve. A moderate increase (to ~ 7 mW/cm^2) results then in a corresponding increase of the current (position 'B'). A further increase, however, invokes tetravalent dissolution at position 'C' behind the current maximum. Finally, it is possible to decrease the photocurrent below its initial value at position 'D' by increase of the intensity to ~ 40 mW/cm^2. Except for small statistical variations, the I-V-curves attain comparable maximum values for intensities > 7 mW/cm^2. This observation is in agreement with earlier studies of the intensity dependent photocurrent behavior of Si-electrodes in dilute NH_4F solutions [201]. Light intensity can therefore be regarded as an equivalent to the applied potential: increase of the photon flux acts as an increased *effective potential*. From a technical point of view, manipulation of the light intensity appears advantageous over variation of the potential since light sources are, in many cases, tunable with more sensitivity than potentiostats. An exception to this equivalence will be discussed below. Figure 3-33, right, illustrates schematically the behavior of the BAR signal when the potential at the respective positions 'A', 'B', and 'C' is kept constant during potentiostatic (chronoamperometric) conditioning. Further increase of the reflectance will then occur except for potentials near 'A' and 'B'. For a potential near 'C' and 'D', behind the first current maximum, the electrochemical reaction involves surface oxidation and oxide etching. As described before, parts of the roughened surface are substituted then by silicon oxide (in several oxidation states) and the reflectometer detects seemingly smaller roughness due to the lower index of refraction of the oxide. Subsequently, the oxide is etched, resulting in even lower reflectance. The reflectance behavior between 'B' and 'C' where a steady-state reflectance behavior can be expected will be separately discussed below.

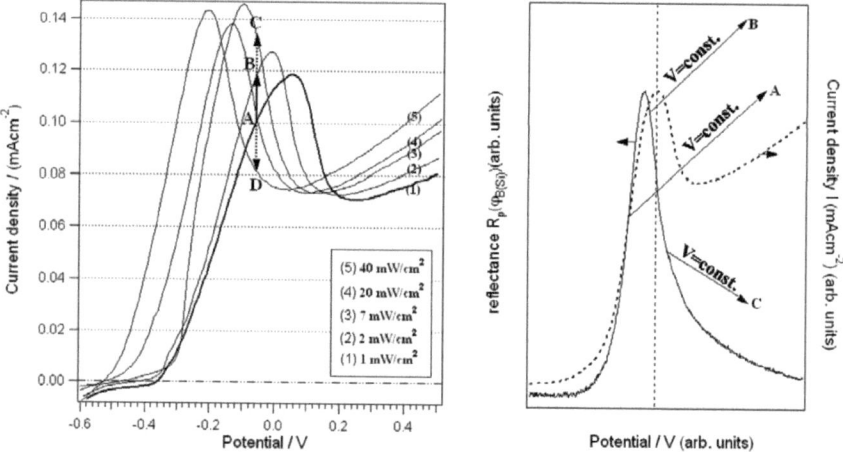

Fig. 3-33: Left: dependence of the photocurrent upon light intensity. At $V = -0.1$ V, arrows indicate the change of electrochemical condition (from A to B, C, and D respectively) when the light intensity is increased. Right: photocurrent and corresponding Brewster angle reflectance data. The behavior of the reflectance for constant potentials, applied at conditions A, B, and C, is schematically shown by arrows.

Fig. 3-34a shows the behavior of the BAR signal with increased light intensities. The phase shift between reflectance and current maxima becomes more distinct with high light intensities. The OCP approaches its final value for intensities > 10 mW/cm^2 (Fig. 3-34b). As fitting curve, shown in Fig. 3-34b, a logarithmic function was chosen. The difference between current and reflectance maximum, in turn, continues to increase, as depicted in Figure 3-34c. Here, a root function was applied to approximate the corresponding behavior.

In Figure 3-34d, the effect of step-modulated light intensity at a low frequency (0.05 Hz) is shown. The sample was roughened in a pre-treatment at -0.1V in order to make the optical response to intermediate oxidation more distinct.

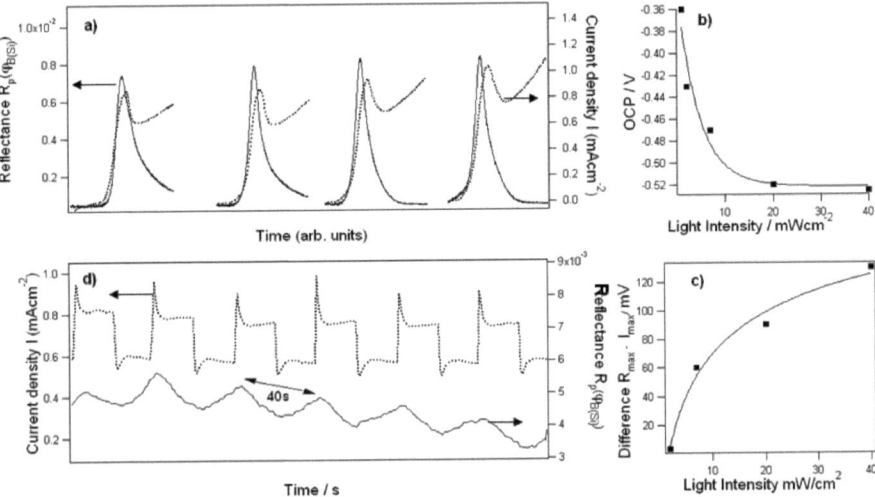

Fig. 3-34: (a) Photocurrents for 1, 2, 20, and 40 mW/cm^{-2} and corresponding BAR reflectance curves. While the OCP shift is limited to ~ -0.52 V (b) the difference between reflectance and current peak shows slower saturation effects (c). (d) Modulated light intensity (20 s per step) leads to alternating divalent and tetravalent electrochemical reactions.

The applied potential, according to the scheme in Fig. 3-33, corresponds to point 'B' using low sample illumination of about 2 mW/cm^2. The photocurrent exhibits relaxation effects, visible as upward and downward peaks, when the intensity is tuned to higher or lower values, respectively. The corresponding change of the surface topography is detected by BAR, indicating alternately surface roughening (and porosity formation) and oxidation (with subsequent etching) of the surface. The phase shift between reflectance and photocurrent indicates a transition of the surface structure in dependence on the light intensity. In order to distinguish the role of potential vs. charge flow, experiments as shown in Fig. 3-35a were carried out for varied scan rates of the potential. In Fig. 3-35, the phase relation between the data was monitored for scan speeds between 1 mV/s and 10 mV/s. The difference between maxima of the reflectance and corresponding maxima of the photocurrent is decreasing with increasing scan rates. Differences of 80 mV, 60 mV, 50 mV and 50 mV were measured for the scan rates of 1 mV/s, 2 mV/s, 5 mV/s and 10 mV/s. This behavior can be similarly described by a logarithmic curve as for increased light intensities shown in Fig. 3-34c. It can therefore be concluded that the surface transformation is primarily caused by the applied potential; the charge flow may contribute to the observation as a term of second order.

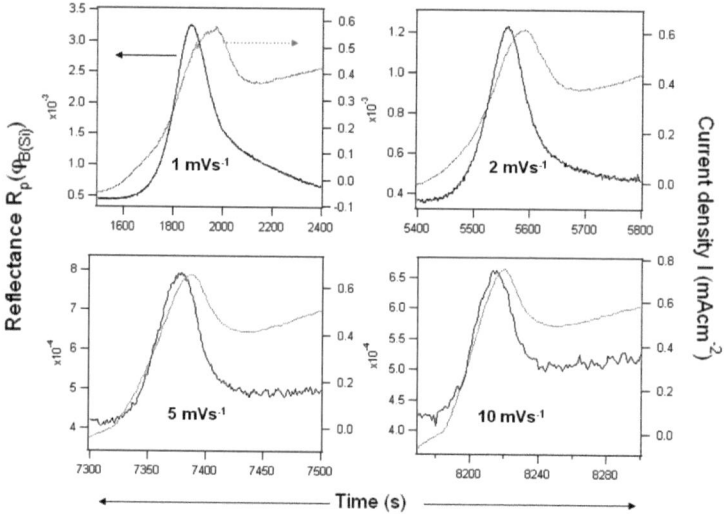

Fig. 3-35: Scan rate variation for silicon dissolution around the first photocurrent maximum and reflectance maximum. The scan rates are indicated in the figure. The corresponding differences between the maxima are 80 mV, 60 mV, 50 mV and 50 mV for increased scan rates.

In order to assess the surface topographies corresponding to point 'A' and 'B' respectively, potentiostatic conditioning was carried out. The treatment at a potential slightly behind the first reflectance maximum (and before the first current maximum corresponding to point 'B') will be discussed first. The potential was scanned until the reflectance passed its maximum (~ -0.1 V) and was held at this value for 15 min as shown in Fig. 3-36a. If surface roughness is assumed to be the only conditioning effect, the slope of the reflectance signal would indicate a decreasing rate of roughness formation. Theoretical calculations, according to the formulae given in section 2.1.2, result in a decrease of the surface roughening rate from 0.4 nm/min in the beginning to 0.2 nm/min at the end of the potentiostatic treatment. The top surface properties were assessed by TM-AFM, and the cross sectional image in Fig. 3-36b proves, in fact, the formation of increased roughness. Cone-like structures on the surfaces with average width of 80-100 nm and average height of 15-25 nm were observed in the corresponding height-mode image. Between the cones, irregular smaller structures are visible. A three-dimensional TM-AFM image of a 500 nm x 500 nm area is shown in Figure 3-36c. The corresponding auto-correlation image is given in Figure 3-35d. Interestingly, a wave-pattern is found that corresponds in its orientation to the distribution of terraces on the originally H-terminated sample as shown in Fig. 3-10. The terraces are aligned, with respect to the primary flat of the silicon wafer (the <110> direction) by nearly 45°.

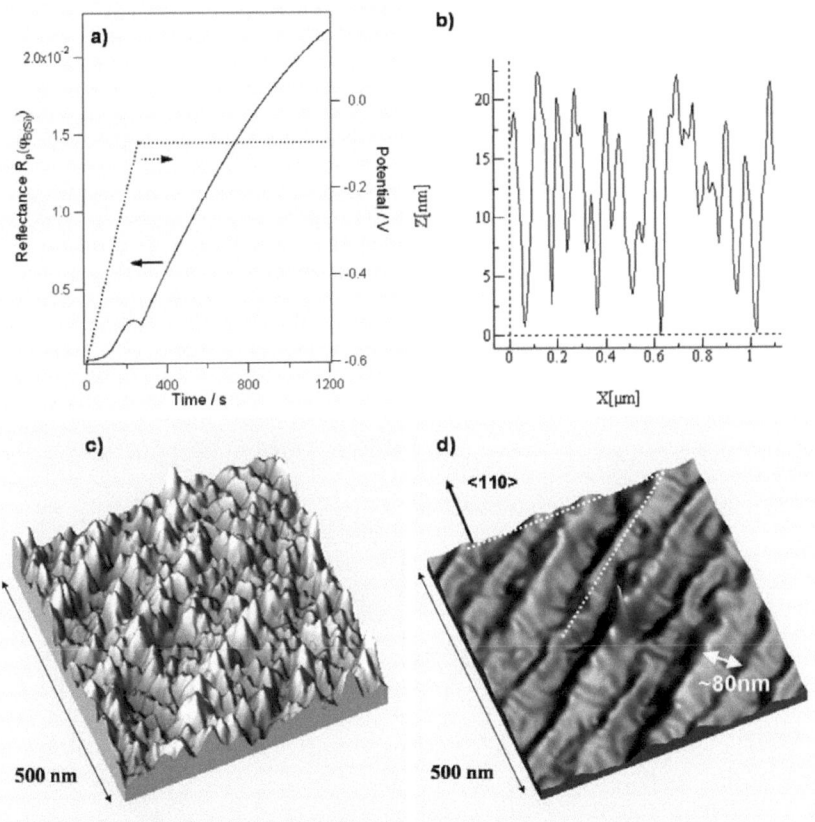

Fig. 3-36: Two-step treatment for nanostructure preparation: i) potential scan to -0.1 V ii) potentiostatic experiment for 15 min. (a) BAR *in situ* data (solid line) and potential (dashed line). (b) Surface profile after preparation. (c) TM-AFM image (height mode); (d) auto-correlation image of the surface shown in (c).

This orientation appears to be preserved in the case of the alignment of the cone-like nanostructures or hillocks. However, while the terraces on the H-terminated sample are separated by about 400 nm, the corresponding distance between the patterns on the electrochemically prepared surface is reduced to 80 nm on average.

The surface topography that corresponds to potentiostatic treatment at point 'A' is shown in Fig. 3-37 as inset number 1. At a potential of V = -0.1 V, the Si surface was roughened for five minutes as indicated by the photocurrent (upper curve) and the *in situ* BAR signal (lower curve) in Fig. 3-37, left. Inspection by TM-AFM shows that the surface exhibits again hillocks but no irregular roughening between the structures.

A subsequent oxidation step was induced by increased light intensity (curve (2) and condition '2' in Fig. 3-37). The surface topography was subsequently assessed by TM-AFM and by High Resolution Scanning Electron Microscopy (HR-SEM). Inset 2 and 2' in Figure 3-37 suggest that the nanostructures are partially removed from the surface. Those structures which are still on the surface appear to be surrounded by a layer presumably consisting of SiO_2. It should be noted that insets 2 and 2' show brighter spots with the same lateral distribution. In the case of HR-SEM the contrast is produced by chemical and topographic information since backscattered electrons were detected.

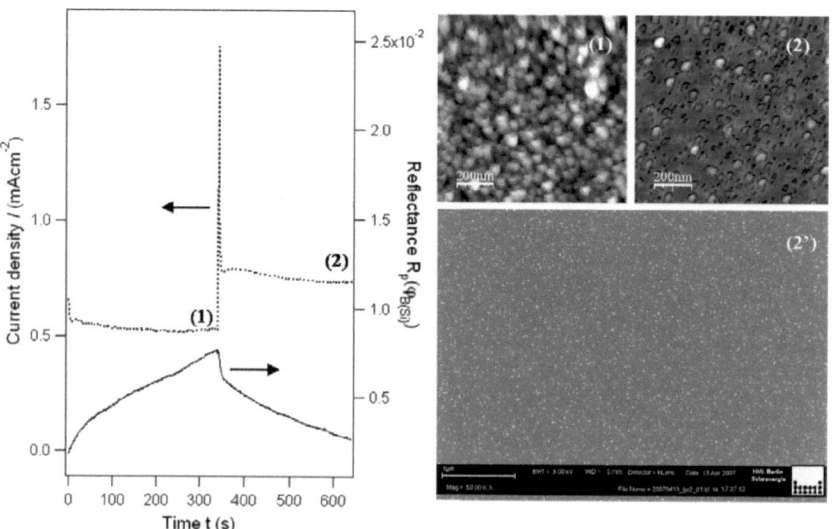

Fig. 3-37: Formation of nanostructures and subsequent oxidation. Left: BAR *in situ* data for lower (1) and higher light intensity (2) and photocurrent curve. Right: TM-AFM image (height mode) before light intensity increase (1) and after experiment (2). Image 2' shows the corresponding result obtained by HR-SEM.

In order to remove the oxide for further AFM analysis, the sample was dipped in 50% HF until a hydrophobic surface condition was attained. Two- and three-dimensional AFM images of all three surface conditions, after nanostructure formation, after photoelectrochemical oxidation and after HF etching are shown in Fig. 3-38. The hillocks are again characterized by alignment effects as visible in part (a) of Fig. 3-38. Moreover, side-walls of individual structures exhibit, to some extent, preferential orientation. With subsequent oxidation and oxide etching, these features are no more visible on the scale of the depicted images (conditions (b) and (c) in Fig. 3-38). The loss of alignment and structure anisotropy would be comprehensible from the induced partial oxidation step which is more isotropic by its nature. However, larger scale surface inspection, to be shown further below, will prove the preserved orientation of the structures with respect to the surface lattice.

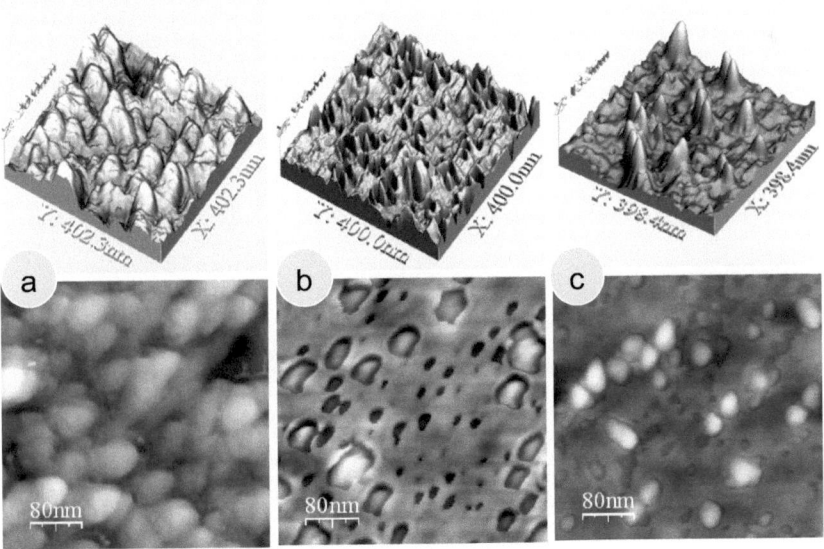

Fig. 3-38: Three- and two-dimensional view of a 400 nm x 400 nm large area after nanostructure formation (a), selective oxidation (b) and etching in 50% HF (c).

A larger survey of the finally obtained surface as well as cross-sectional analysis is shown in Figure 3-39. The TM-AFM image in Fig. 3-39 exhibits silicon nanostructures with a density of about $10^{10}/cm^2$. Heights and widths of the structures are distributed with small deviation from average values. The analysis of the corresponding profiles for the initial condition, the condition after oxidation, and the final condition after the HF dip are shown as insets. While the cone-like structures at condition '1' are 6-10 nm high and ~ 60 nm wide, the structures after oxidation (and oxide removal) show a small increase in height and a distinct decrease in width. The overall gain in the aspect ratio is in the range of 40-60 %.

Alignment effects of the structure distribution are stressed by dashed lines in both, the AFM height-mode image (left) and the calculated autocorrelation image (right). On the scale of these images the preserved initial topography of the terraced Si(111) sample with miscut angle towards $<1\bar{1}2>$ is clearly recognizable.

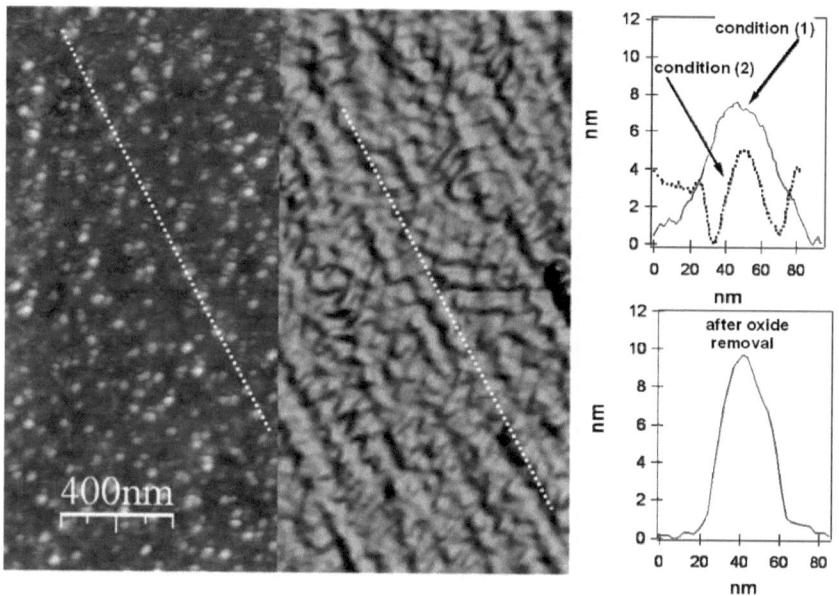

Fig. 3-39: Si surface after removal of the oxide by HF. Left: TM-AFM image. Right: nanostructure profiles before oxidation, after oxidation and after oxide removal.

In order to analyze the dependence of nanostructure alignment on the surface lattice topography, Si(111) samples with customized miscut angle towards <011> were prepared accordingly.

The result is shown in Fig. 3-40, right. By increase of the contrast, the difference in the structure distribution is clearly visible. Surface height analysis by AFM shows that individual structures are about twice as high as those obtained on the surface shown in Fig. 3-38. These observations correspond to increased photocurrent densities and a larger slope of the BAR signal although experimental conditions, as light intensity and applied initial potential were chosen in accurate agreements to the conditions related to Figs. 3-37 through 3-39. For comparison, the complete surface topography of Fig. 3-39 is shown again.

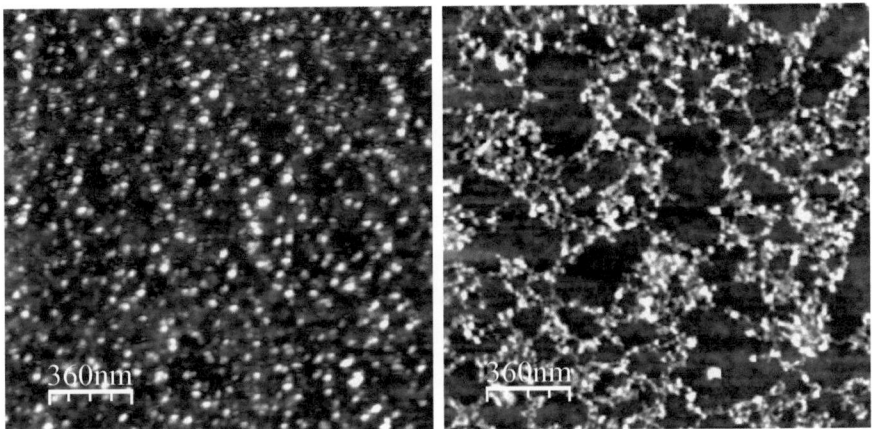

Fig. 3-40: Nanotopographies prepared on Si(111) by subsequent divalent and tetravalent reaction steps. Left: miscut angle towards $<11\bar{2}>$, 0.5°. Right: miscut angle towards <011>, 4°. The contrast of the images was increased. The average height of the structures, shown on the right-hand image, is twice as large as on the left-hand image.

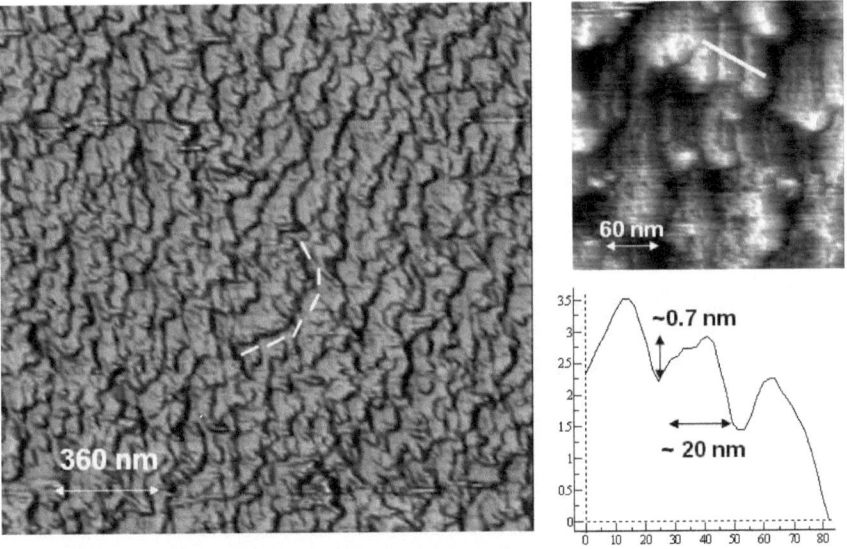

Fig. 3-41: CM-AFM height image of the Si(111) sample with 4° miscut angle towards <011>. The left-hand image is a top-view of a three-dimensional visualization that suppresses the fine-structure but emphasizes the curved terrace geometries (see dashed curve). The fine structure, in turn, is depicted on the right-hand side in the upper image. Small atomic steps are visible with step width of about 20 nm and step height of multiples of one bi-layer (3.14 Å).

The surface topography of the sample (Fig. 3-40, right) exhibits nanostructures, aligned in ring-like networks. The area between the structures is rather flat. An rms value of about 1 nm was determined by AFM. In order to investigate the relationship of nanostructure alignment effects to the initial surface topography after chemical H-termination, a sample with 4° miscut angle towards <011> was etched for 100 s and subsequently 2 min in NH_4F (40%). Larger etching times resulted in strong surface corrugation, indicating that the dissolution rate is markedly increased by the miscut angle. This observation also accounts for the increased photocurrent densities observed during electrochemical preparation of the nanotopographies in Fig. 3-40. Inspection of the chemically prepared surface proved irregularly distributed terraces. These terraces are shown in Fig. 3-41, left, by the top-view of a three-dimensional representation. This view was chosen to emphasize the curvature of the terraces only. The fine-structure is depicted in an individual image on the right-hand side. Cross-sectional analysis shows small areas of about 20 nm length with step height of multiples of the typical bi-layer distance of Si(111) terraces (3.14 Å). Comparison of Fig. 3-40, left, and Fig. 3-41, right, clearly proves a dependence of the lateral nanostructure distribution on the underlying surface lattice properties. A tentative interpretation will be given in the following section.

The obvious shaping effect of the oxidation step, induced by increased photon flux towards the surface, is finally investigated. In the following experiment, a *constant-reflectance method* was developed: after an initial potential scan close to the first reflectance maximum, the light intensity was varied in order to keep the reflectance value constant. The change in light intensity should serve for small variations of the *effective potential*, leading thus to alternating nanostructure formation and oxidation steps. If nanostructure formation, and therefore surface roughening, would be the only effect during divalent dissolution near the reflectance maximum (while the light intensity is varied to keep the optical response constant), the mean roughness should not change significantly because an increasing mean thickness of the roughness layer would also increase the reflectance.

The experimental principle is schematically shown in Fig. 3-42. The reflectance maximum is indicated as R_{max}, the corresponding current maximum, according to the literature [65], as J_{PS}. Here, the abbreviation 'PS' refers to the term 'porous silicon'. Behind J_{PS}, according to the accepted interpretation, electropolishing takes place, i.e. the surface condition is characterized by a steady-state relationship of anodic oxidation and etching. Since the anodic oxide layer forms a thin, smooth and dense oxide layer, a smoothening of the surface is expected. Porous silicon formation is not observed in this region.

The circle in Fig. 3-42 indicates the applied potential while the curved arrow represents the change of the *effective potential* invoked by variation of the light intensity.

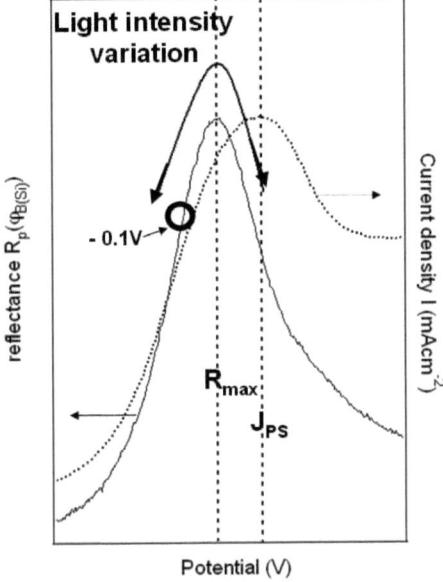

Fig. 3-42: Scheme for nanostructure manipulation at constant potential (- 0.1 V) and varied light intensities (variations were in the 10 $\mu W/cm^2$ range). By variation of the photon flux, the effective potential moves along the range indicated by the arrow. Surface corrugation is expected to increase/decrease accordingly.

In Fig. 3-43, reflectance and photocurrent behavior are shown for constant light intensity during the first 1400 s and for the following period of 1000 s where the light intensity was alternately adjusted to higher and, respectively, lower values to impede a further increase of the reflectance. The grey bar in Fig. 3-43 indicates the small fluctuations of the reflectance during the second period of the experiment. The variation of the light intensity was very small, in the range of approximately a few 10 $\mu W/cm^2$ according to the indicator on the power supply which was calibrated in a preceding investigation. This effect corresponds to the close alignment of the curves in Fig. 3-33 where only a few mW/cm^2 separate divalent from tetravalent dissolution reaction schemes (compare points 'B' and 'C').

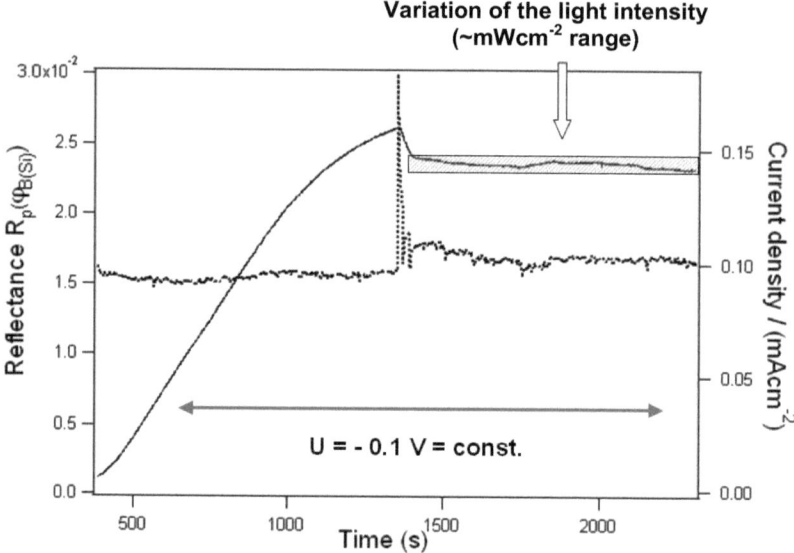

Fig. 3-43: Reflectance and photocurrent behavior for constant light intensity (until 1400 s) and for varied light intensity (between 1400 and 2400 s). The grayed bar indicates the reflectance variation during the second period of the experiment. The applied potential was throughout –0.1 V.

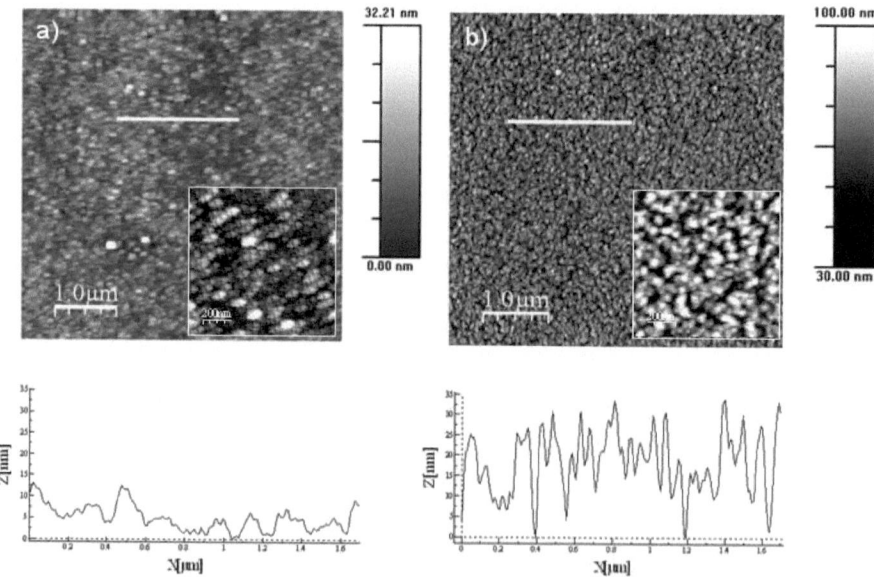

Fig. 3-44: TM-AFM surface and cross-sectional analysis after 1400 s and after light intensity variation for further 1000 s in the *constant-reflectance mode*. The rms changes from 3.6 nm to 8.4 nm. A magnification of the topographical surface condition is presented as insets.

In Fig. 3-44, two TM-AFM images are presented which correspond to the surface state after the initial period of about 1400 s (Fig. 3-44a), where the reflectance signal shows a continuous increase, and to the final surface state after the *constant-reflectance* treatment (Fig. 3-44b). Although the constant reflectance signal suggests no significant change at the surface, a pronounced roughness increase is observed after photoelectrochemical conditioning. This result is a clear indication of sub-surface modification of the silicon sample with topographic features probably far below the wavelength of the probing light (500 nm). It is therefore assumed that microporous silicon has formed with increasing porosity, i.e. with increasing ratio of the respective volumes of nanometer-seized cavities and bulk material.

3.2.2 Model considerations for the self-organized and engineered nanostructure formation

The discussion of the findings, presented in the previous sections, has to address four observations in separate considerations:

1. The formation principle of nanostructures protruding out of the silicon surface or indented in the surface (porosity formation),
2. The observed shift of the current-voltage curve upon illumination of n-type silicon and
3. The effect of selective oxidation on pre-nanostructured Si-surfaces.

From Figs. 3-34a and 3-35 clearly follows that the first reflectance peak precedes the corresponding first current peak during dissolution of silicon in 0.1 M NH_4F *pH* 4. The voltage range, relative to the OCP, is characterized by the transition from divalent to tetravalent dissolution. This transition can be assumed not to evolve instantaneously as a sharp step function but rather to develop in a more continuous way.

For solutions of hydrofluoric acid, the dissolution valence was determined in dependence of the anodic current density upon galvanostatic treatment (see Fig. 3-45) [202]. J_{PS} in this image denotes the so-called critical current density which corresponds to the first current peak in the current-voltage characteristics. The two curves refer to p-Si, measured in the dark, and strongly illuminated low doped n-Si, respectively. The change in the dissolution valence from 2 to 4 is furthermore accompanied by a transition to a diffusion limited reaction, i.e., if the rate of anodic oxide formation increases, the transport of the oxidizing species to the silicon interface is hindered. According to Fig. 3-45, this limitation can be approximately attributed to the range where divalent and tetravalent processes contribute in equal parts to the dissolution.

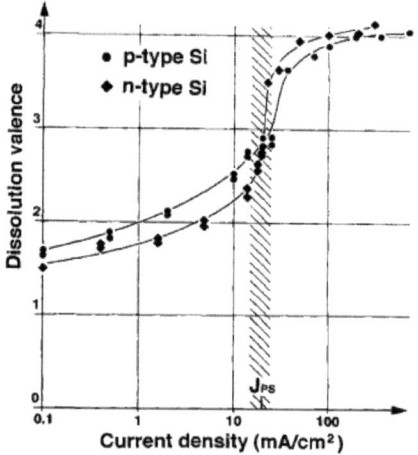

Fig. 3-45: Dissolution valence for n- and p-type silicon under galvanostatic conditions. The relation for p-type silicon was obtained for dark current measurements. The corresponding relation for n-type silicon was obtained under strong illumination. The image was taken from [202].

By light intensity variation, described in section 3.2.1, a constant reflectance was achieved in a range where the dissolution valence is probably between 2 and 3. Larger tetravalent dissolution rates resulted in the formation of pronounced subsurface porosity and the effective dielectric constant decreased to a degree that the effect of increasing surface roughness was compensated (in terms of optical analysis). Therefore, the optical response remained almost unchanged. This interpretation is supported by the known behavior of silicon in the divalent dissolution regime where porous silicon formation can be expected.

A further (topographic) transition can thereby be deduced which accompanies the change in the chemical reaction during increase of the dissolution valency. The chemical transition was earlier described by Lehmann and Föll [202, 203]: the process of silicon dissolution in HF containing solutions can be schematically divided in intervals which are separated by the points of inflection and, respectively, maximum points that are visible in the current density behavior. The corresponding points in the reflectance characteristics almost perfectly coincide with this scheme as shown in Fig. 3-46. Here, current density and reflectance are shown around the first peak together with the corresponding first derivatives. The important points of the current density are numbered from 1 to 3, corresponding points of the reflectance are indicated by capital letters, A to D. According to the interpretation by Lehmann, point 1, the first point of inflection, indicates commencing oxidation that limits for the first time the increase of the current density. Point 2, the first current maximum was observed to represent the threshold value for porous silicon formation. Point 3 finally indicates the formation of a thin but dense oxide layer that suppresses the current flow. The experimental findings described above suggest now that point A, which coincides with point 1 of the current, indicates beginning and increasing formation of nanopores, i.e. the commencement of a topographic transition.

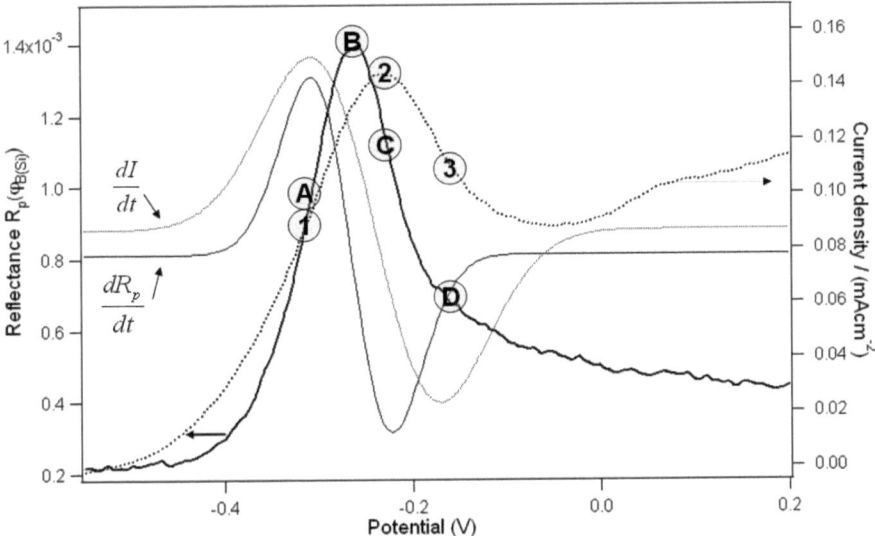

Fig. 3-46: Schematic representation of the potential dependent dissolution behavior of n-type Si(111) electrodes in 0.1 M NH$_4$F solution of *p*H 4. Indicated by the respective arrows, the photocurrent density and the reflectance are shown. Additionally, the corresponding derivatives are depicted. Points 1 through 3 and A through D are discussed in the text.

This interpretation makes it comprehensible that the slope of the reflectance becomes less steep since the porous silicon layer, beneath the continuously roughening surface, is characterized by a smaller dielectric function. Point B, i.e. the reflectance maximum, is caused by a porosity that probably exceeds a threshold value, as discussed in section 2.2.2, leading thus to a decrease of the optical response upon further dissolution. The surface layer is characterized by an effective medium that, although expanding in depth, attenuates the reflectance response due to increased porosity. From point C on, the decrease of the reflectance becomes less steep since surface oxides in a sub-monolayer range give rise to additional reflection. A pronounced change of the reflectance behavior is observable from point D on where the slower reflectance decay is due to dominant oxidation before the chemical-topographic conditioning leads towards the electropolishing regime.

Topographic analysis for surfaces prepared at potentials at and below point A of the curve in Fig. 3-44 showed that cone- or hillock-like structures are forming. From optical analysis, discussed above, follows that sub-surface porosity formation is presumably of minor importance. The arising structures were proven to be dependent on the atomic step density as well as the miscut angle of the substrates. The results of Figs. 3-36 through 3-41 suggest not only an anisotropic effect in the formation scheme. Since structures are observed at sites corresponding to terrace edges, the anisotropy appears *reversed*. While the chemical dissolution in NH_4F preferentially removes surface atoms located at the edges, flat terraces at a distance from the edges seem to be dissolved at a higher rate during photoelectrochemical conditioning. As a result, remnants of these edges are directly observable in AFM measurements or, at least, after pattern intensification by an auto-correlation procedure. As a working hypothesis for future investigations, variations of the specific termination (by H, F or OH) of silicon atoms at edges and on flat terraces can be considered. As indicated by UPS measurements, silicon dissolution in NH_4F gives rise to surface dipole formation (in contrast to the corresponding treatment in HF). Surface dipoles, in turn, lead to a perturbation of the surface potential. The interface cannot be regarded as equipotential area in this case. This might locally increase and, respectively, decrease the dissolution rates. Further experiments are necessary to provide an in-depth analysis on the atomic level.

Oxide formation, employed on pre-nanostructured surfaces by light intensity variation, is expected to proceed isotropically. The isotropy, however, can be disturbed by nanotopographies with sharp edges which locally increase the electric field distribution. Furthermore, silicon atoms exposed to the expanding SiO_2 volume may be more susceptible to the proceeding oxidation process due to stress effects. For that reason, selective oxidation of rough areas between nanostructures as well as of the structure side-walls, as observed by the experiments in Figs. 3-38 through and 3-40, may evolve according to the scheme, depicted in Fig. 3-47.

It is assumed that rough areas will be preferentially oxidized upon transition from the divalent to the tetravalent reaction pathway. Growing oxides will expand and approach the nanostructures in the vicinity. This possibly leads to compressive stress in both the oxide and the nanostructure. Stress, in turn, will facilitate the oxidation of the nanostructure side-walls. A phenomenon, that eventually leads to the observed increase of the aspect ratio in Fig. 3-38.

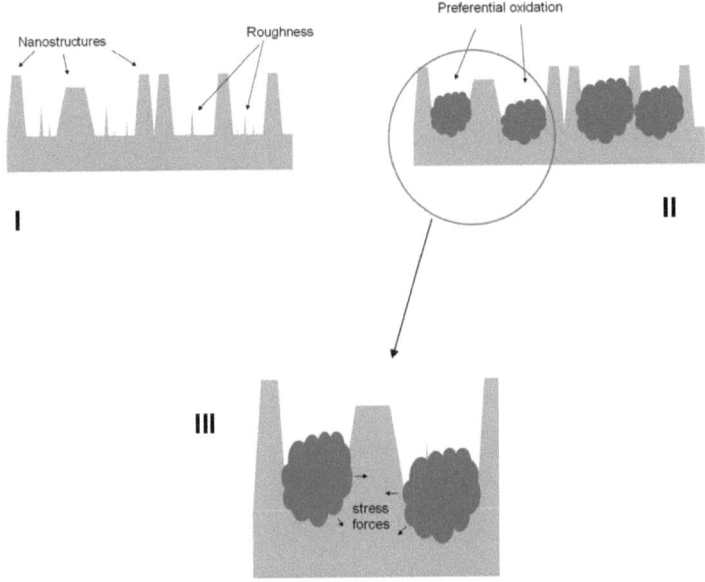

Fig. 3-47: Model consideration of selective oxidation of pre-structured Si(111) surfaces. I: nanostructured surfaces are forming during chronoamperometric treatment; II: light induced increase of the effective potential leads to selective oxidation at sharp edges; III: oxide induced stress leads to dissolution of the side-walls of the nanostructures.

Since oxidation should also take place on the top of the nanostructures, only a fraction of them will pertain to exist. In future experiments, limits for the nanostructure aspect ratios have to be determined.

The shift of the current-voltage characteristics, exploited by selective oxidation of pre-nanostructured Si(111) surfaces, can be assigned to two contributions of different physical nature. Firstly, the shift of the OCP can be discussed in terms of the Butler-Vollmer equation [53], developed for non-corrosive reactions. According to this equation, the anodic current, as part of the exchange current, will increase upon increased photon flux. If a linear increase with light intensity can be assumed, then the OCP can be calculated by:

$$0 = A\Phi_0 \exp(C_1 V) + B \exp(-C_2 V) \tag{3-8}$$

where A, B, C_1 and C_2 are determined by the respective concentration, the transmission factor and temperature. V denotes the applied potential. Rearrangement of Eq. 8 yields:

$$-\frac{B}{A\Phi_0} = \exp\left[V(C_1 + C_2)\right] \text{ and}$$

$$\frac{1}{(C_1 + C_2)} \ln\left(-\frac{B}{A\Phi_0}\right) = \frac{-1}{(C_1 + C_2)} \ln\left(\frac{A\Phi_0}{B}\right) = V_{OCP} \tag{3-9}$$

i.e., a logarithmic law. In fact, the OCP variation in Fig. 3-34 shows an almost perfect logarithmic behavior and a curve of this type was used for the approximation.

The OCP for p-Si behaves differently upon increased illumination. The increase of cathodic contributions to the exchange current is too low in order to shift the OCP larger than 20 mV. For larger cathodic potentials, however, the current density increases significantly. Furthermore, the current density maximum is fixed to its position in the dark. This is illustrated in Fig. 3-48.

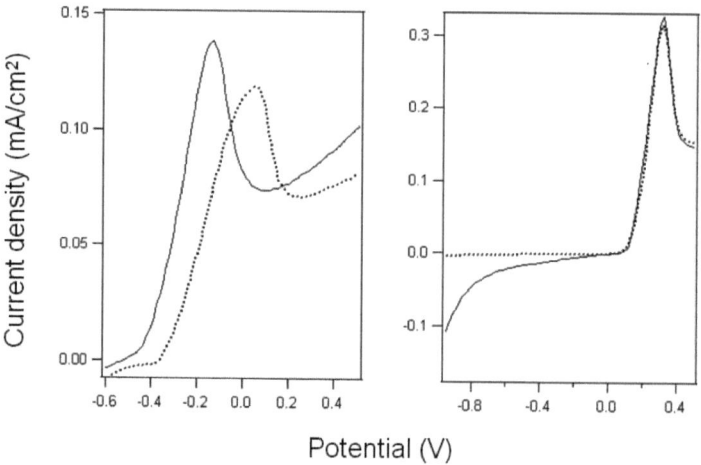

Fig. 3-48: Response of n- and p-type silicon electrodes to (increased) illumination. While n-type electrodes are characterized by a cathodic shift of the current-voltage characteristics, p-type electrodes show only a small shift of the OCP (~ 20 meV).

Since the maximum photocurrent values of n-type silicon increase only by a small magnitude, charge carrier supply appears not to be an important limitation in this voltage range. With respect to the complexity of the reaction scheme around the first current maximum, explanation of the current-voltage shift is speculative on the basis of the results obtained so far. For some types of electrodes, minority carrier caption by surface states upon illumination is well known [161, 162]. Such process can lead to a shift of the Fermi-level and unpinning of the band-edges. But a detailed discussion requires knowledge of the distribution of surface states, i.e. capacity measurements by, for instance, impedance spectroscopy. Therefore, further investigations have to clarify the role of surface states, the flat band position upon illumination and, generally, the charge distribution at the silicon/electrolyte interface.

3.2.3 Structural changes at the Si(111) interface during anodic oscillations

In order to investigate if regular surface topographies on silicon are forming during photocurrent oscillation cycles and if these topographies are related to the observed oxide thickness variation, the relationship between growing anodic oxide layers and the interface structure was investigated. *In situ* BAR can be regarded as indicator for the modification at the surface-electrolyte interface as already shown in the preceding sections. For photocurrent oscillations, however, the silicon interface is constantly buried beneath the oxide layer and the assessment is therefore even more intricate than in the divalent dissolution range.

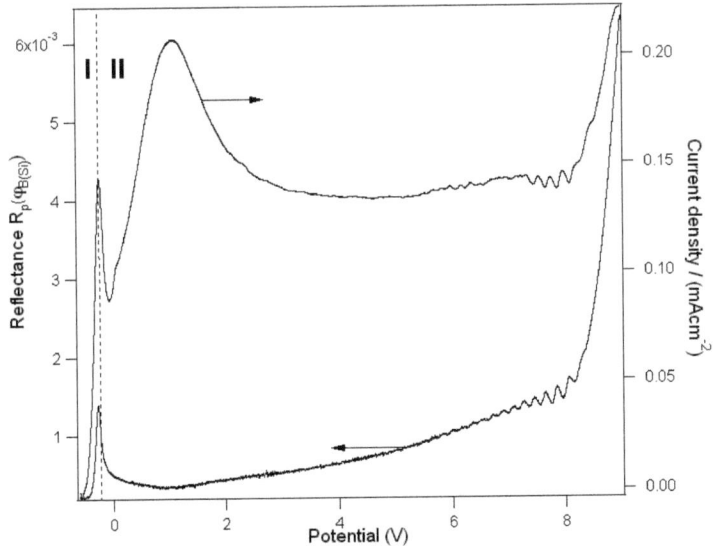

Fig. 3-49: Photocurrent characteristics of n-Si(111) (upper curve) and *in situ* BAR reflectance signal (lower curve). Regions of divalent (I) and tetravalent (II) dissolution are separated by a dashed line.

At low scan rates, the optical response to the surface conditioning process of silicon photoelectrodes is more distinct than at higher scan rates as already shown in Fig. 3-35. This follows from the larger charge flow, consumed in the reaction, and therefore the larger volume of the roughened/porous/oxidized layer on the top of the substrate. Photocurrent oscillations at low potential scan rates are likewise more pronounced as shown in Fig. 3-49. In this experiment, the scan rate was set to 2 mV/s. Oscillation cycles in potentiodynamic experiments are characterized by a continuously increasing average oxide layer. This can be concluded from the BAR reflectance oscillations in Fig. 3-49 which are superimposed to an almost linear increase of the signal.

However, quantitative analysis of these effects necessitates information on oxide thickness variation and on interface roughness evolution. BAR (BAA) responds equally to increasing oxide thickness and surface/interface roughness. Quantification of these parameters demands optical analysis over an extended spectral range or direct assessment of surface/interface topographies by AFM measurements. In aqueous environments, a particular sensitivity of *in situ* BAR to interface roughness can be deduced by the simulation shown in Fig. 2-5. Due to the similarity of the respective dielectric constants, ε_A (of the electrolyte) and ε_{ox} (of the oxide), changes of R_p during oxide growth are smaller than during roughness variations at the oxide/silicon interface. This follows from the fact that the effective dielectric constant $<\varepsilon>$ of the rough layer, modeled by Bruggeman effective medium approximation, is generally larger than ε_{ox}. Roughness changes at the interface therefore strongly influence the monitored reflectance signal.

In order to assess the interface condition during photo-oscillation cycles, photocurrent oscillations were monitored at constant potential of U = 6 V as shown in Fig. 3-50. Current maxima and those of the reflectance are observed at the same frequency but separated by a constant phase shift. A small average decrease of the reflectance signal is accompanied by decreasing maximum current values.

For a first assessment of the monitored signal, it can be assumed that the whole charge flow contributes to the modification of the SiO_2/Si system including roughening of the interface. If there is no electrochemical etching of the oxide, current and reflectance can be related to each other by:

$$R_p \approx A + B \int_0^t I_{photo} dt . \qquad (3\text{-}10)$$

Eq. 3-10 expresses that there exist parameters A and B such that the reflectance can be represented by a linear transformation of the charge flow measured in an experiment during the time interval [0, t]. If chemical etching takes place, i.e. an additional process without current flow, Eq. 3-10 has to be modified as carried out below.

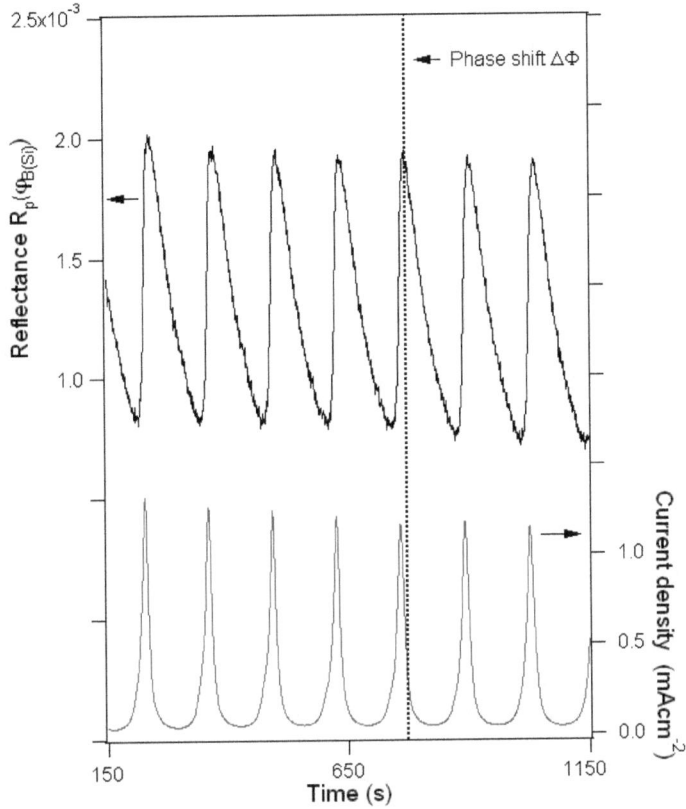

Fig.3-50: Photocurrent oscillations of n-Si(111) at a constant potential of 6 V. A vertical line is shown to indicate the phase shift between BAR reflectance (solid line) and current density oscillations (dashed line).

In this case, parameters A, B and β, γ exist such that the reflectance and the charge can be related to each other by:

$$R_p(\varphi_B(SiO_2)) = A + B\int_{t_0}^{t}(I_{photo}(t) - B\gamma)dt = A + B(Q_{photo}(t) - Q_{photo}(t_0)) - B\gamma t. \quad (3\text{-}11)$$

Here, γ represents the chemical etch rate that attenuates the measured (and calculated) reflectance). In other words $\beta \cdot \gamma \cdot t$ represents the charge that is not any longer 'stored' in the oxide layer. The parameter B is dependent on the dissolution valence (v = 4), the number of surface atoms per cm² and the dielectric function of the oxide. The parameter A determines any offset due to the reflectance of the initial surface condition. In Fig. 3-51, measured reflectance values and those calculated by Eq. 3-11 are compared.

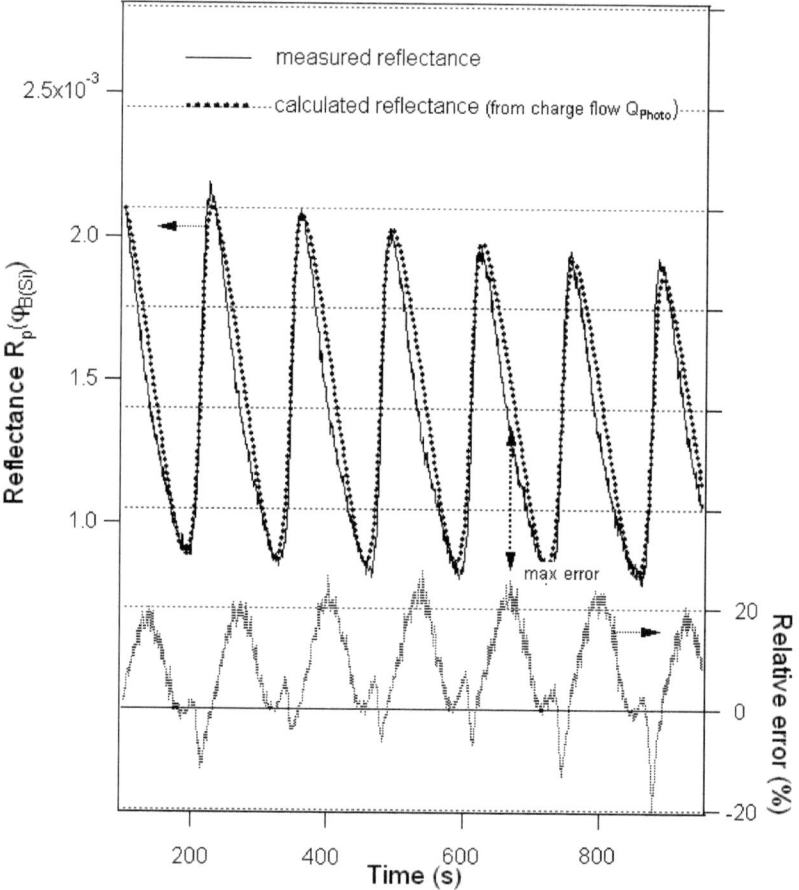

Fig.3-51: Relationship between the oscillating photocurrent and a linear transformation of the charge flow according to Eq. 3-11.

The calculation was carried out to achieve best agreement between local maxima and minima. Therefore, no least-square fit was applied. Phase behavior and local maxima are well simulated.

The ascending parts, however, coincide better with experimental values than the descending slopes. These deviations may result from either varying oxide etch rates within a single oscillation cycle, changes in the surface topography of the oxide layer or corresponding changes at the oxide/silicon interface. While varying oxide etch rates, resulting from local pH variation, are difficult to assess, corresponding changes at the surface and interface can be investigated by AFM analysis. For that purpose, the oscillation cycles were interrupted at the presumed minimum and, respectively, maximum of the oxide layer thickness, and the oxide morphology was analyzed by AFM. Subsequently, BAA was applied during repeated etching steps in NH$_4$F (40%) according to the procedure in section 3.1.2 in order to determine the oxide thickness and to carefully expose the interface.

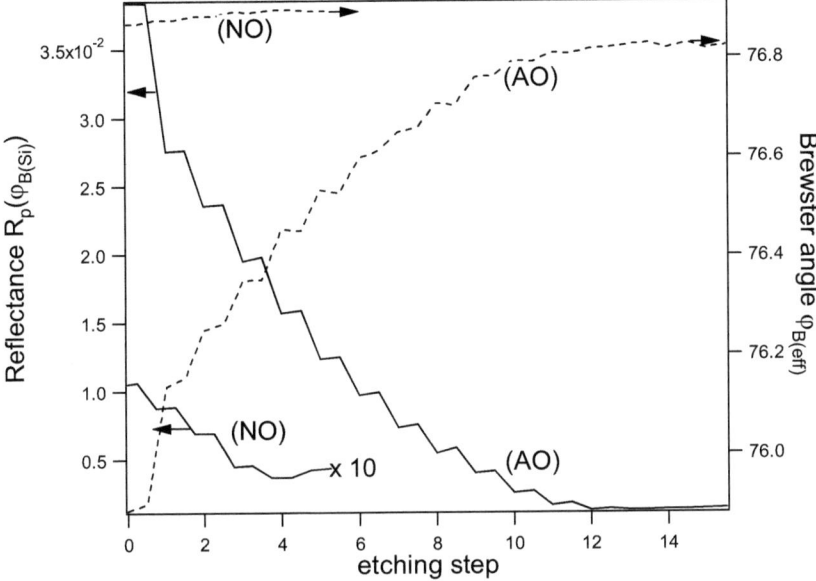

Fig. 3-52: Measured Brewster angle (dashed line) and reflectance (solid line) at this angle. (AO): during 14 etching steps of the anodic oxide in 40% NH$_4$F, 10 s each; (NO): during 5 etching steps of the native oxide in 40% NH$_4$F, 20 s each. The reflectance ('NO') is shown after magnification by a factor of 10.

In Fig. 3-52, the etching steps for maximum oxide thickness are shown in comparison to the values obtained during the etch-back of a native oxide layer. Corresponding to the larger oxide thickness, reflectance values are distinctively increased at the early stage of etching while Brewster-angle values are lower by about 1°. After 12 etching steps, employed for 10 s each, no pronounced change in reflectance or Brewster-angle position are recognizable. Lowest $R_p(\varphi_B)$ values are obtained for step 13. They are larger than the initial values on the native oxide indicating roughening (see below). Therefore any additional roughening behavior as observed on native oxide covered samples during etch-back of the oxide in 40% NH$_4$F is not visible in this case. Fig.3-53 shows TM-AFM data of the anodic oxide before etching steps were applied. 1 μm x 1 μm images are exhibiting columnar islands with diameters of 50-80 nm.

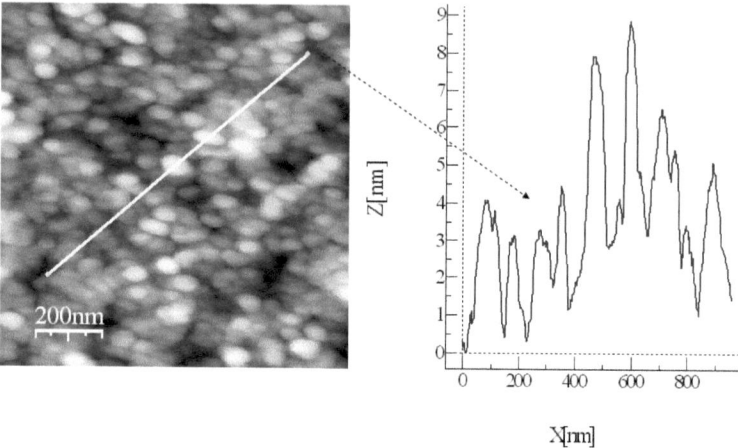

Fig. 3-53: Surface analysis by TM-AFM for maximum oxide thickness during anodic oscillations. (a) Height-image on a 1 μm x 1 μm area. (b) Cross sectional analysis along the white line, indicated on the left-hand side.

The same procedure was applied for the corresponding minimum oxide thickness, i.e. the silicon sample was removed from the electrolyte at the minimum of the reflectance. Fig. 3-54 shows the decrease of the reflectance during 7 etching steps until a minimum reflectance value is obtained.

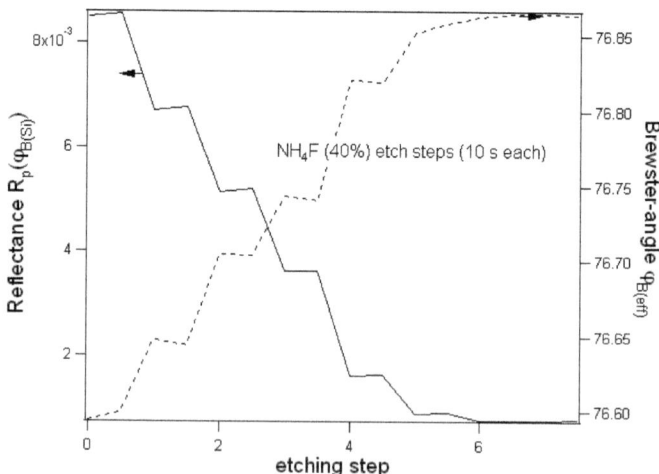

Fig. 3-54: Measured Brewster angle (dashed line) and reflectance (solid line) at this angle during 7 etching steps, 10 s each, of the minimum anodic oxide layer in 40% NH_4F.

TM-AFM analysis reveals for the surface before etching that the columnar oxide growth is characterized by larger oxide islands of about 200 nm width:

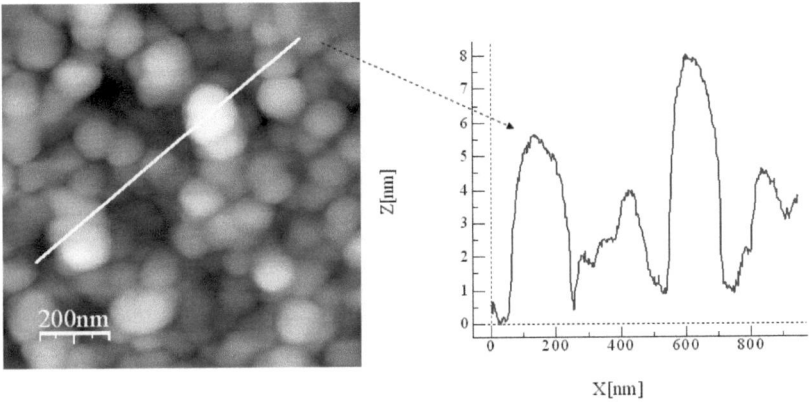

Fig. 3-55: Surface analysis by TM-AFM for minimum oxide thickness during anodic oscillations: (a) Height-image on a 1 µm x 1 µm area. (b) Cross sectional analysis along the white line as indicated on the left-hand side.

In Fig. 3-56, the interface for both states of the oscillation cycle (oxide minimum and oxide maximum, respectively) after exposure to NH_4F etching is shown.

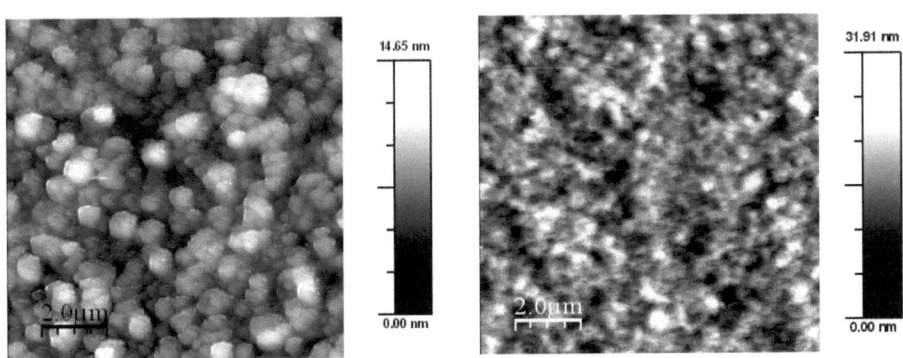

Fig. 3-56: Surface analysis by TM-AFM of the SiO_2/Si interface after NH_4F etching of the oxide layers with minimum oxide thickness (left) and maximum oxide thickness (right).

More important than differences in the respective root mean square roughness (1 nm compared to 1.9 nm) appears the phenomenon of microstructures at the interface, originally buried beneath the anodic oxide layer of minimum thickness. The structures appear to be of almost uniform shape and size (~ 500 nm). Increased interface roughness at times of maximum oxide layer thickness, as depicted in Fig. 3-56, right, can be understood by magnification of the respective AFM images.

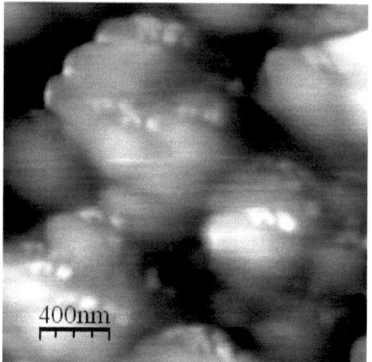 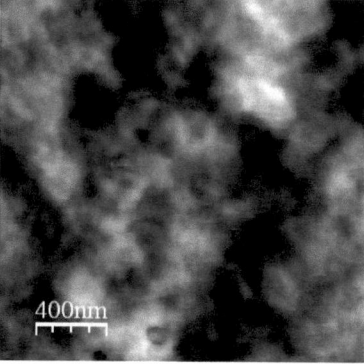

Fig. 3-57: Magnification of the interface topographies shown in Fig. 3-56. Surface analysis by TM-AFM of the SiO_2/Si interface after NH_4F etching in the case of minimum oxide thickness (left) and maximum oxide thickness (right) during anodic oscillations.

The structured topography, observed for minimum oxide thickness, appears to be destroyed but is still recognizable. It can be concluded that the microstructures are attacked by an isotropic dissolution mechanism leaving behind side-walls with increased porosity. This effect may be attributed to the larger stress forces beneath the thicker oxide which result in more irregular interface oxidation and the observed porosity formation.

Anisotropic interface topographies as in Fig. 3-57, left, were also observed for Si(100). In this case, square holes were observed after removal of the anodic oxide layer by HF [204].

Finally, the information provided by optical and microscopic investigations were used to determine the respective oxide thicknesses on the basis of the known surface and interface morphologies shown in the preceding images. This calculation can serve, in principle, for calibration of the parameters A, B and β, γ in Eq. 3-11 and are useful for comparative analysis of the charge transfer across the semiconductor-electrolyte interface.

The surface height profiles in Figs. 3-53 and 3-55 were used to determine void-oxide fractions for subsequent EMA analysis. Exemplifying for this approach, Fig. 3-58 shows the calculation of the void-oxide distribution for the oxide layer related to the maximum thickness. The standard height deviation σ was calculated resulting in the corresponding rms value. The thickness of the roughened layer is given by $2|\sigma|$.

The value for the volume fraction of the silicon oxide host material is given, on the other hand, by the integral of the height profile between the ± σ lines divided by the total area. As a result, the rough layer is almost equally composed of oxide and voids.

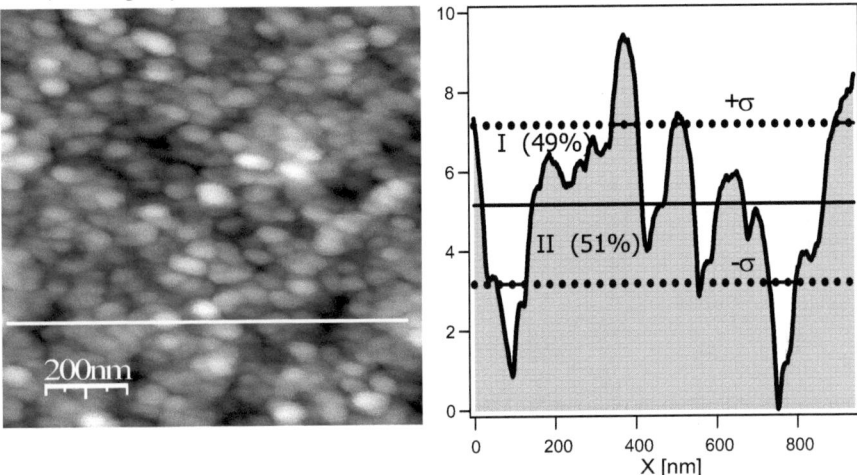

Fig. 3-58: Left: TM-AFM height image of the silicon oxide surface at maximum oxide thickness. The distance between white and black colors corresponds to 12 nm. Right: Surface profile, according to the white line on the left-hand side, indicating the mean height (solid horizontal line) and standard deviations ±σ (dashed horizontal lines). σ represents the root mean square value (rms) obtained by AFM roughness analysis. The different media used in optical multi-layer analysis (ambient and silicon respectively) are denoted by the roman numerals I and II. The respective volume fractions of silicon and oxide enclosed by ±σ are 0.51 and 0.49 respectively.

Since the investigated multi-layer structure consists of ambient, ambient/anodic oxide, anodic oxide, anodic oxide/silicon and silicon, detailed AFM analysis was also performed for the respective anodic oxide/silicon interfaces. The measurements show that the roughness at the ambient/anodic oxide interface varies by 20% around a base value of 2 nm. At the anodic oxide/silicon interface, rms values are changing from 1 nm (minimum oxide thickness) to 1.9 nm (maximum).

For numerical analysis, the following parameters were used: (i) ambient/anodic oxide interface, rms = 2, void fraction v = 0.5; (ii) anodic oxide/silicon interface with rms = 1.9 (oxide maximum) and rms = 1 nm (oxide minimum), void fraction v = 0.5. As result, a refractive index n_{ox} = 1.38 for the oxide and a thickness variation between 4.8 nm and 13.8 nm were calculated. n_{ox} is ~ 5% smaller than published values for thermal oxides. This result is related to the open micro- and nanostructure of the oxide, its fast etching rate and its softness in contact-mode AFM experiments. As discussed in chapter 1, the presence of water inclusions lowers considerably the oxide quality. The determined etch rates, 0.1 nms^{-1} for the anodic oxide, and 0.01 nms^{-1} for the native oxide, give support to this interpretation.

Interestingly, the phenomenon of roughness evolution during photocurrent oscillation cycles can be associated with the surface state density at this interface. Repeated oscillation cycles were reported to result in an improved interface quality with higher photoluminescence yield and lower surface recombination [205]. The analysis carried out here confirms that repeated oscillations improve the interface quality by lowering the interface roughness. However, intermediate dark current transients have to be applied in order to achieve this improvement. This result follows from Fig. 3-59 where the previously determined parameters A, B and β, γ, used for the simulation in Fig. 3-51, were also applied to earlier oscillation cycles of the same sample.

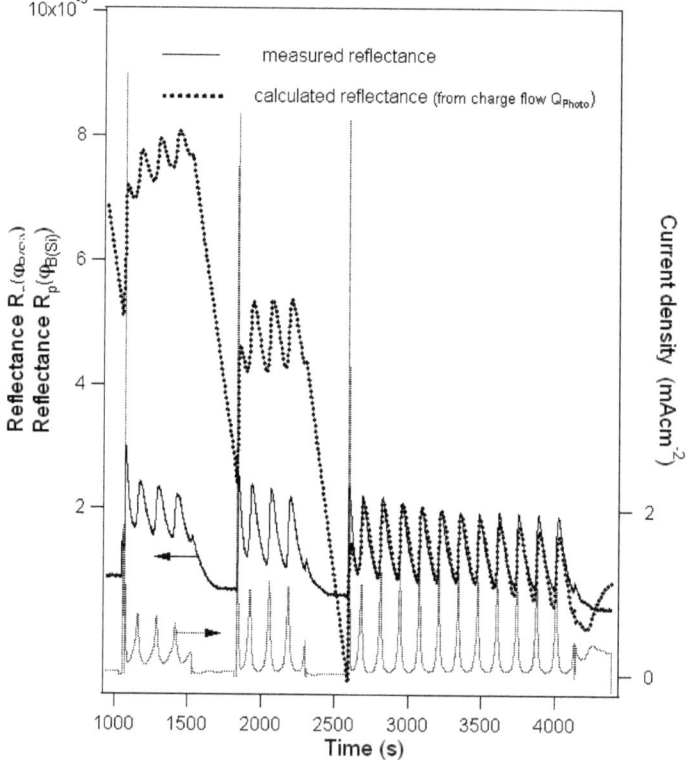

Fig. 3-59: Application of the parameters A, B and β, γ in Eq. 3-11 to earlier photocurrent oscillation periods, separated by intermediate dark current transient treatment. The approximation fails for earlier periods, i.e. charge flow, surface and interface roughness are changing markedly during repeated periods with intermediate oxide removal.

It can be seen that during earlier periods the approximation according to Eq. 3-11 fails, i.e. the current flow was considerably lower while the corresponding reflectance shows an offset that can be attributed to increased interface roughness. In order to match the respective curves, individual parameters have to be chosen. These parameters characterize therefore the respective charge transfer processes as well as the changing surface-interface condition.

Although micro- and nanostructuring effects are observable at the oxide-silicon interface, systematic manipulation of these structures appears difficult. The interface can only be manipulated by electrolyte and light intensity variation. Moreover, real-time assessment and control by BAR is impeded since the interface topographies are buried beneath the oxide layer. The optical signal, in this case, provides information averaged across the respective interfaces and the oxide layer. All these quantities interfere in an unpredictable manner and require post-experimental analysis by selective etching and AFM interface investigation. For these reasons, anodic oxide oscillations were not employed for further structuring experiments.

3.2.4 Summary: *in situ* controlled self-organized nanostructure formation

Silicon nanostructures prepared in the region of commencing divalent dissolution were produced under control of *in situ* BAR. Two novel conditioning techniques were thereby developed to manipulate these pre-conditioned surface topographies:

1) Shaping of the nanostructures in the tetravalent dissolution region by variation of the light intensity corresponding to an increased *effective potential*.
2) Formation of sub-surface porous layers with increased porosity by application of the *constant-reflectance method*.

Nanostructures are topographically characterized by hillock-like geometries protruding out of the surface. Comparison between surfaces with specified miscut angle towards the $<11\bar{2}>$ and <011>-direction showed a distinct relationship between near-surface lattice properties and structure alignment. These findings complement earlier reports of isotropic silicon nanocrystal formation on (100)-oriented surfaces in HF and NH_4F, respectively [193-195]. The anisotropies in shape and lateral distribution suggest likewise anisotropies in the electrochemical dissolution reaction as recently discussed [206]. Detailed analysis of the structure geometries necessitates in the future high-resolution analysis by, for instance, Scanning Tunneling Microscopy (STM). These measurements, however, have to be carried out in vacuum or in an inert gas atmosphere in order to avoid unwanted surface oxidation.

In the oscillatory regime, a simplified equation was derived in order to relate charge flow and BAR reflectance. This relation accounts for the observed phase shift between photocurrent and reflectance oscillations. Furthermore, the assessment of changing interface conditions was shown to benefit from this mathematical approach.

3.3 Self-organized propagation of fractal silicon microstructures in concentrated NH₄F

3.3.1 Background on fractals and macropores

Fractal geometries are well known in mathematics since ancient times: the Apollonian circle, for instance, exemplifies the process of pattern recurrence on a circular template in which tangent triple circles are inscribed (see Fig. 3-60). Free areas are then infinitely filled by so-called *inner Soddy circles* with decreasing diameters to achieve the final form [207]. Although the self-similarity is not perfect as in typical fractal images: scaling effects are well exemplified, i.e. the increasing number of elemental figures with decreasing size. With the famous question of Benoit Mandelbrot: "How long is the British coastline?" [208], fractal geometries came into focus of a broader audience and the discussion of a fractional dimension, which generalizes the corresponding term of the Euclidean geometry, was investigated on virtual objects such as Julia and Mandelbrot sets [209] as well as on phenomena in biology, economy, chemistry and physics [210-212].

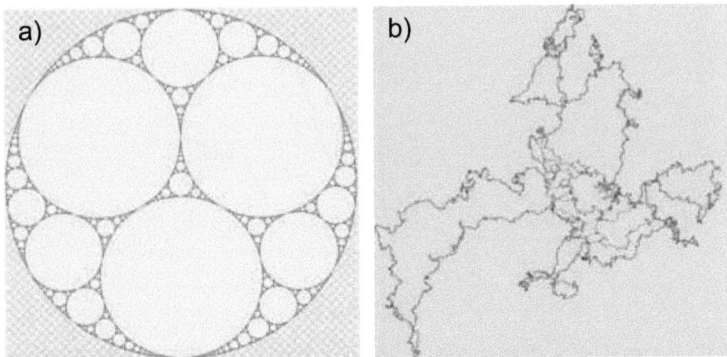

Fig. 3-60: Basic fractal geometries: (a) Apollonian circle which exemplifies a construction scheme of circles infinitely inscribed in a circular boundary. (b) Coast line of a Brownian archipelago that has fractional dimension.

As for silicon, fractal analysis of surface roughness phenomena [213, 214], fractal interpretation of porous silicon formation [215, 216] or pattern recurrence during (electro-) deposition processes were discussed [217]. In the following, the accelerated dissolution process of silicon (photo-)electrodes in concentrated ammonium fluoride is analyzed which shows many aspects of fractal geometries such as pattern recurrence and scaling effects.

Furthermore, the pronounced self-organization of the photoelectrochemical system is investigated with respect to the feedback mechanism that is most responsible for the observation of regular or aperiodic structures. The relation to macropore formation [218] will be discussed in the final section (3.3.6).

3.3.2 Experimental results for Si(111), (110), (100) and (113) surfaces

The (photo-)electrochemical dissolution of silicon in NH$_4$F (40%) shows specific deviations in the current-voltage behavior compared to the dissolution in diluted solutions of pH 4 (see section 3.2). The dark current, as shown in Fig. 3-61, initially increases from OCP (~ -1 V) and shows an almost uniform plateau of ~ 200 µA/cm^{-2} from about 0 V on. Hydrogen evolution was observed throughout, suggesting electron injection at a high rate in the course of silicon dissolution [57].

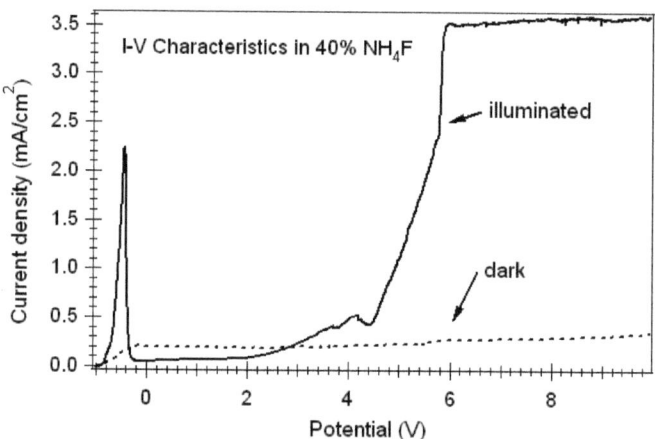

Fig. 3-61: Current-voltage characteristics of n-Si(111) in the range of -1 V to 10 V, scan velocity 10 mV/s. Solid curve: with illumination (light intensity 5 mW/cm^2); dashed curve: measurement in the dark.

For comparison, the dark current in diluted solutions (0.1 M, pH 4) does not exceed about 1 µA/cm^2. Under illumination (full curve), a first maximum current peak at U = - 0.5 V, near OCP, indicates divalent silicon dissolution. From 4 V on, a distinct current increase is observable while gas bubbles are adsorbing to specific surface sites. In this range both increased silicon oxide formation and oxygen evolution reaction (OER) are expected. For potentials > 6 V, the steady-state current indicates a dynamic equilibrium between anodic oxide formation, oxide etching and oxygen evolution as reported recently for the p-Si/acetate (borate) buffer system [219]. Inspection of the surface by optical microscopy revealed microscopic traces across the whole electrode area for experiments carried out in the dark as well as under illumination. The optical microscope was equipped with a polarizing optical unit and a camera system. In Fig. 3-62, one half each of the electrode area is shown. The images are composed from smaller single micrographs.

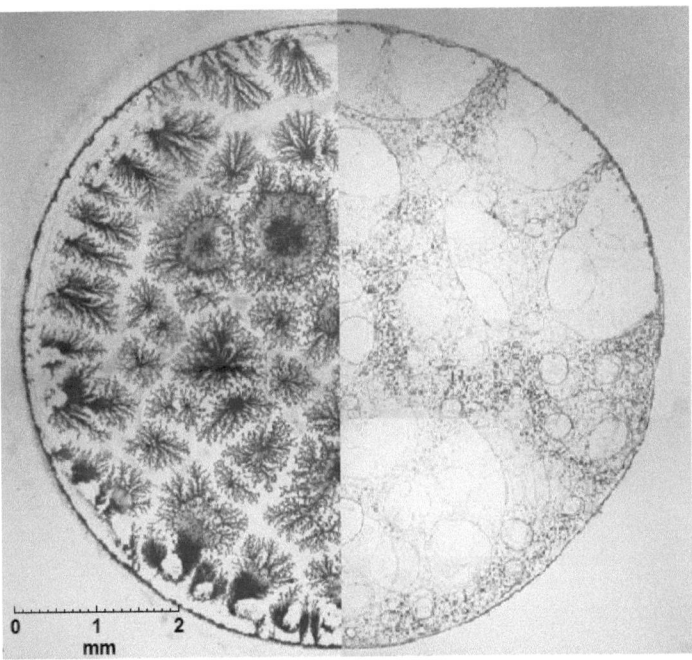

Fig. 3-62: Optical micrograph of the n-Si(111) surface after potential scan from -1 V to 10 V with illumination (left) and in the dark (right). Ramifying dendrites are visible allover the left area. The outer circular boundary results from the VITON O-Ring used in the electrochemical cell.

Without illumination (Fig. 3-62, right), remnant circular etch lines are present on the surface indicating the positions where hydrogen bubbles have been adsorbed to. Comparison to the Fig. 3-60a shows similarity to the construction principle of the Apollonian circle. Deep etch triangles are surrounding the hydrogen bubble sites and are visible as dark spots. For the illuminated surface, dendritic formation of etch grooves, subdivided in a number of domain-like structures, appears as the predominant feature (Fig. 3-62, left). Except for the area close to the O-Ring, radial growth of the dendrites can be observed. These grooves are spreading from defective areas where oxygen bubbles have been observed during the photoelectrochemical treatment. Dendrites belonging to different domains appear to repel each other.

In Fig. 3-63, the time-dependent current density behavior of a Si(111) sample is shown for the potential U = 6 and 2.75 V. The current keeps almost constant such that potentiostatic and galvanostatic treatments represent similar procedures.

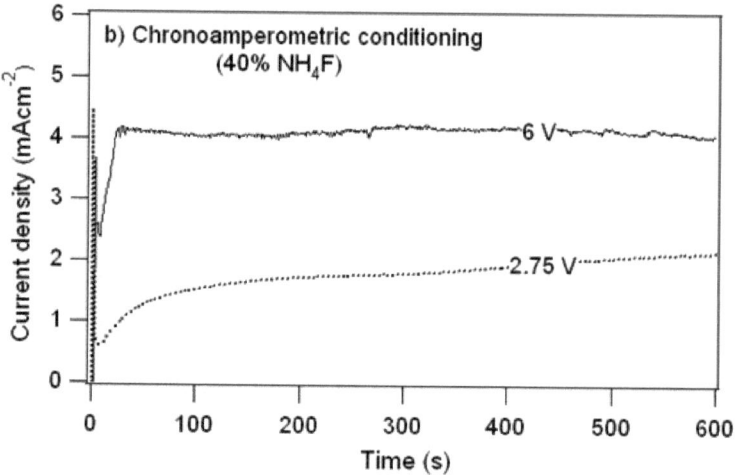

Fig. 3-63: Potentiostatic treatment of n-type Si(111) in NH$_4$F (40%) at U = 6 and 2.75V, respectively. The light intensity was about 7 mW/cm^2.

Variation of the light intensity at 6 V showed an almost linear dependence between the photon flux and resulting photocurrents. This relationship holds true over a wide range of intensities from about 2 mWcm^{-2} to 60 mWcm^{-2} (not shown here).

The electrode surface, investigated by SEM, showed after preparation at 6 V regular structures as shown in Fig. 3-64. The Si(111) etch structure exhibits a nearly perfect hexagonal boundary surrounding an almost unstructured circular center where the initially adsorbed oxygen was located. The view onto the whole sample area revealed again ramifying structures but only near the O-Ring where shadowing effects reduce the light intensity and are promoting dendrite formation. The inset of Fig. 3-64a exhibits a complex ensemble of deeply etched microfacets arranged in repeated segments of similar geometry. A preferential orientation of the facets can not be concluded from this image. Even more perfection and agreement with the corresponding surface lattice is found for Si(100) surfaces (Fig. 3-64b). While some of the branches are still separated, others have combined to build an almost complete square.

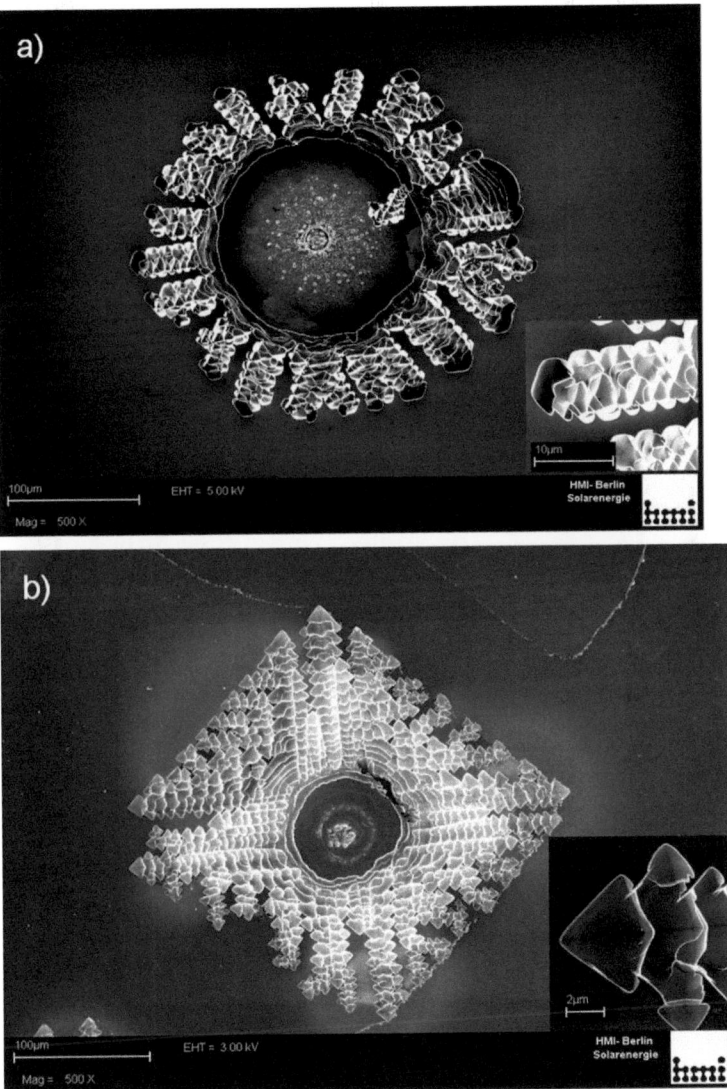

Fig. 3-64: Regular pattern formation obtained by potentiostatic treatment (6V) for 10 min. (a) Hexagonal structure on n-Si(111). (b) Square structure on n-Si(100). The insets show the respective equal-sized microfacets which are aligned along the propagation direction of the structures.

According to the inset in Fig. 3-64b, side-walls of the microfacets enclose an angle close to 45° with respect to the propagation direction. It can be therefore assumed that the microfacets are oriented along the <111> direction as expected for alkaline dissolution of Si(100) substrates. Anodization of Si(111) and Si(100) samples for only one minute showed that the center circular areas, which are attributed to immediate oxygen evolution, are rapidly built. At the boundaries, initial cracks can be observed. Corresponding observations were made during chemical etching in NH_4F (40%): in this case hydrogen evolution initiated crack formation as well as porous layer formation beneath the bubbles [220]. The simultaneous presence of a reactive gas, an aqueous liquid and the bare silicon substrate results obviously in an increased dissolution process at the phase boundary. In Fig. 3-65, the evolution of the oxygen bubble on the silicon surface, maybe covered by an ultrathin oxide layer which allows electron tunneling, is shown as observed by the experiment (left) and as a schematic (right).

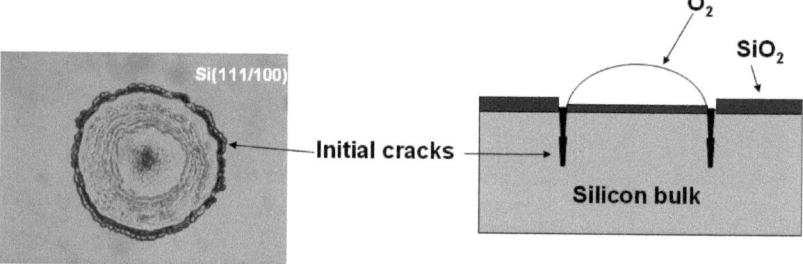

Fig. 3-65: Evolution of gaseous oxygen and adsorption to the silicon surface in the initial phase of silicon dissolution in NH_4F (40%) at a potential U = 6 V. Left: experimental observation with optical microscopy after 1 min. Right: schematic drawing of the adsorption process and initiation of deep cracks around the circular adsorption area.

Oxygen evolution in alkaline solutions is thought to proceed in the presence of OH^- molecules, adsorbed to the surface. The overall reaction is [53]:

$$4\ OH^- \rightarrow O_2 + 2\ H_2O + 4\ e^-. \tag{3-12}$$

This scheme involves the following steps:

1. $4\left(OH^- \rightarrow OH_{ad} + e^-\right)$
2. $2\left(2OH_{ad} \rightarrow H_2O + O_{ad}\right)$
3. $2OH_{ad} \rightarrow O_2$

Oxidation of the silicon surface as well as oxide etching, which are the main reactions here, is accompanied by H^+ evolution. At high oxidation rates, local *pH* variation may be therefore induced. The anodic oxygen reaction scheme of Eq. 3-12 has then to be exchanged for the overall reaction

$$2 H_2O \rightarrow O_2 + 4 H^+ + 4 e^-. \qquad (3\text{-}13)$$

This reaction involves:

1. $4\left(H_2O \rightarrow OH_{ad} + H^+ + e^-\right)$
2. $2\left(2OH_{ad} \rightarrow H_2O + O_{ad}\right)$
3. $2OH_{ad} \rightarrow O_2$

For a first assessment of the local dissolution rate, SEM images as shown in Fig. 3-64 are not sufficient. They reveal the lateral component of the dissolution rate only, which is extremely high. Structures with an extent of about 200 µm evolve during 10 min of anodization. Compared to the chemical dissolution of silicon in NH₄F (40%), as discussed in section 3.1.2, the lateral component of the dissolution rate is more than 4000 times larger. In order to assess the vertical component, profilometry measurements were carried out for a structure formed on Si(100).

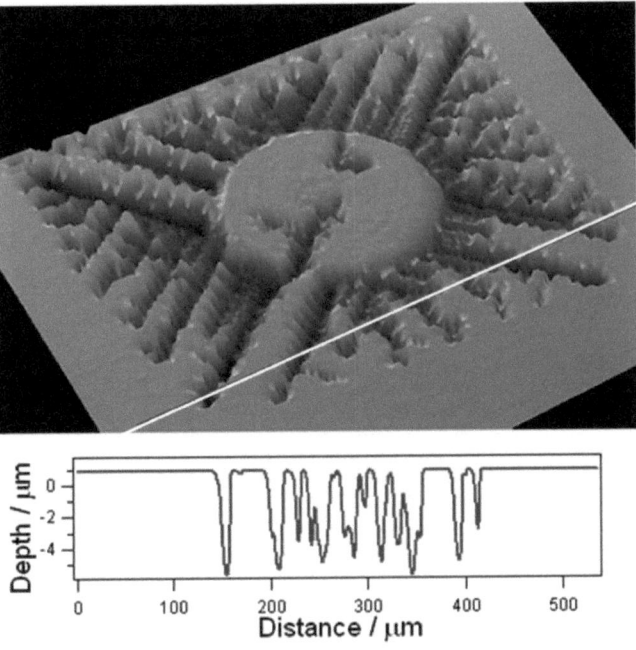

Fig. 3-66: Profilometry analysis of an etch structure on Si(100). The Si electrode was anodized at U = 6 V for 10 min. Illumination intensity was about 7 mW/cm². Maximum depths of about 6 µm were measured by cross sectional analysis shown below.

The image is shown in Fig. 3-66 with the corresponding cross section. This cross section indicates that the dissolution towards the bulk is about 0.6 µm per minute. Therefore lateral and vertical dissolution rates probably differ by at least one order of magnitude.

At lower potentials of 2.75 V, as shown in Fig. 3-67, traces of adsorbed oxygen bubbles are not visible.

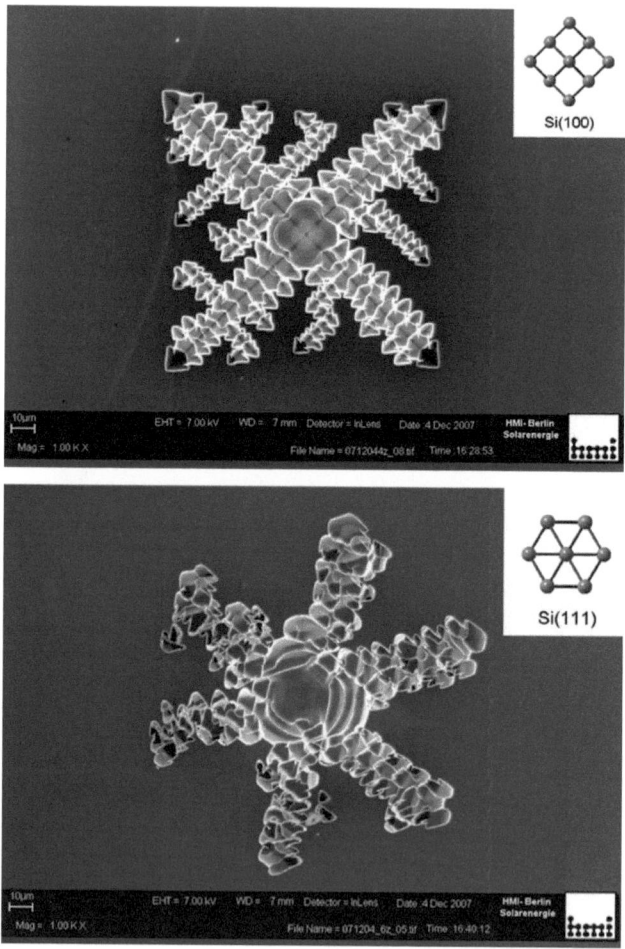

Fig. 3-67: Fractal structures on Si(100) and Si(111) as indicated by the corresponding insets. The structures were investigated by SEM after 10 min etching time in 40% NH_4F at a potential of U = 2.75 V. The light intensity was about 7 mW/cm^2.

However, structures are forming with even more perfect symmetry compared to Figs. 3-64a and b. Particularly, the hexagonal symmetry on Si(111) is less disturbed by the center circular area, caused by oxygen adsorption, as seen in Fig. 3-64a. The branches are well separated from each other and exhibit dendritic growth. Although the circular areas, related to initial oxygen bubbles, are not visible in Fig. 3-67, etch structures were observed on other samples (under high illumination) which suggest that also in this case small gas bubbles are forming and are initiating the growth of the structures. In Fig. 3-68, a CM-AFM image and a corresponding SEM image are shown, revealing small square structures on a Si(100) surface with a semispherical inner topography which presumably has to be attributed to initial oxygen adsorption and subsequent increased silicon dissolution.

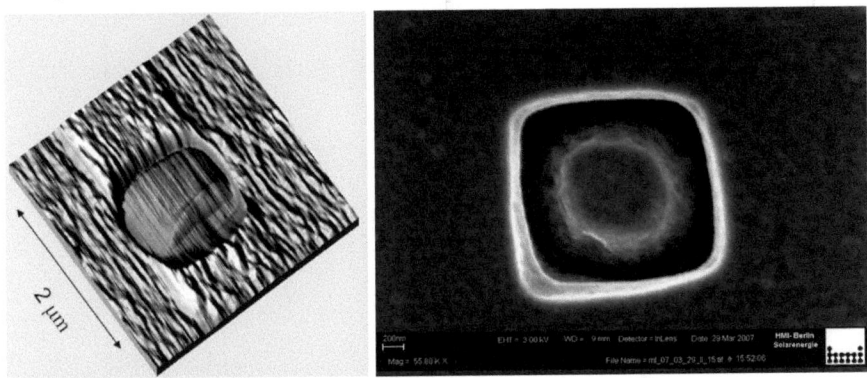

Fig. 3-68: Initial square structures on Si(100) obtained after 10 min etching in 40% NH_4F at a potential U = 6 V. The light intensity was 20 mW/cm^2. Left: CM-AFM image. Right: SEM image of a corresponding structure.

At low potentials (U = 1 V), no etch structures were observed. By comparison with the current-voltage characteristics in Fig. 3-61, increasing photocurrents, to be related to oxygen evolution, are obviously required to initiate and drive forward the structure growth. Even initial cracks, produced in separate experiments at 6 V, did not propagate at U = 1 V. In order to determine the chemical surface state for these experimental conditions, SRPES was carried out after surface preparation for 10 min at 1 V and 2.75 V, respectively. The survey spectra showed especially high oxygen and fluoride signals. The investigation of the Si 2p core level signal, shown in Fig. 3-69, proves a decreasing ratio of the photoelectron yield for the silicon bulk signal (at $E_B \sim 99.5$ eV) and the corresponding SiO_2 signal at $E_B > 104$ eV with increasing anodic potential. Quantitative analysis results in the respective oxide layer thickness of 1.4 nm (U = 1 V) and 2.8 nm (U = 2.75 V). Already for the 2.8 nm thick oxide layer, charging was observed due to the high brilliance of the synchrotron light source. Two dashed lines, shown in Fig. 3-69, indicate the resulting decrease of the measured kinetic energies of about 1 eV.

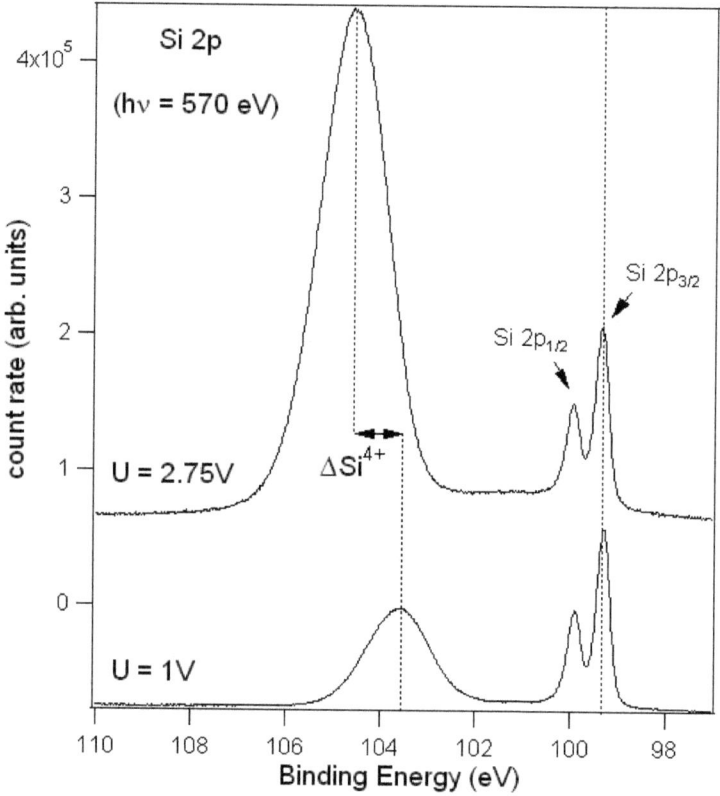

Fig. 3-69: SRPES analysis of the Si 2p core level after preparation of Si(100) photoelectrodes in 40% NH$_4$F at U = 1 V (lower curve) and U = 2.75 V (upper curve). Oxide charging resulted in a shift of the binding energy of the Si^{4+} signal for the thicker oxide layer as indicated by the dashed line.

The fluorine F 1s signal was analyzed with varied excitation energies, ranging from 750 eV to 1250 eV, allowing thus the determination of a depth profile of the fluoride distribution across the oxidized surface. The corresponding mean elastic scattering length of photoelectrons changes thereby from below 1 nm to a few nm depending on the homogeneity of the anodic oxide. The results for the Si(100) sample, prepared at U = 1 V and U = 2.75 V, are shown in Fig. 3-70a and b. The curves were normalized and corrected, if necessary, for any charging induced shift of the binding energies. A constant offset to the normalized count number was added in order to separate the curves more clearly.

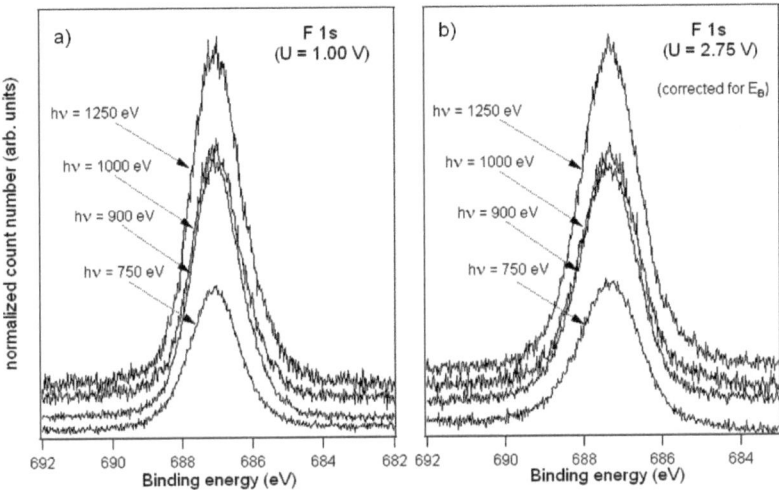

Fig. 3-70: SRPES analysis of the F 1s signal with varied excitation energies after preparation of Si(100) photoelectrodes in 40% NH_4F at U = 1 V (left) and U = 2.75 V (right). Charging effects resulted in a shift of the binding energy which was corrected in the right image.

Although remnant fluorides are known to cover the silicon surface after preparation in NH_4F containing solutions, the signals shown in Fig. 3-70 are very pronounced and were well reproducible in subsequent experiments. The signal increase with increasing excitation energy suggests that the F^- species migrate through the oxide layer and are preferentially located at the SiO_2/Si interface. The ability of the F^- anions to pass through the oxide layer is increased by the inhomogeneous structure of the anodic oxide which was discussed earlier in section 3.2. Further proof of this interpretation will be given below where the oxide structure of backside illuminated Si(111) will be discussed by TM-AFM analysis.

The high dissolution rate of p-type silicon and n-type silicon with excess doping levels ($> 10^{17}$ cm^{-3}) in NH_4F containing solutions impedes the controlled formation of surface topographies with high regularity. In Fig. 3-71, the corresponding current-voltage curves, measured in the dark, are shown. Although the reaction mechanism differs for the two electrodes (hole extraction in the case of p-Si and electron injection in the case of (n+)- Si), the curves are clearly related to each other: the lower doping density of the p-type electrode (10^{15} cm^{-3})

results in the same features as observed for the highly doped n-type electrode. For p-Si, the rapid decrease of the current density at U1 = 0.7V and U2 = 3.8V is due to the formation of an insulating anodic oxide layer. With continued potential increase, promoted by oxide etching, migration of reactive ions starts again due to the increased electric fields across the layer. This behavior resembles the observation of the so-called Flade-potential on metal electrodes. While on metal electrodes the formation of an insulating layer is observed only once (any subsequent current increase results from oxygen evolution), the corresponding layer formation on the silicon electrode occurs at least two times (behind OCP and near 3 V). For potentials larger than 5 V, oxygen evolution may take place rather than further oxide formation.

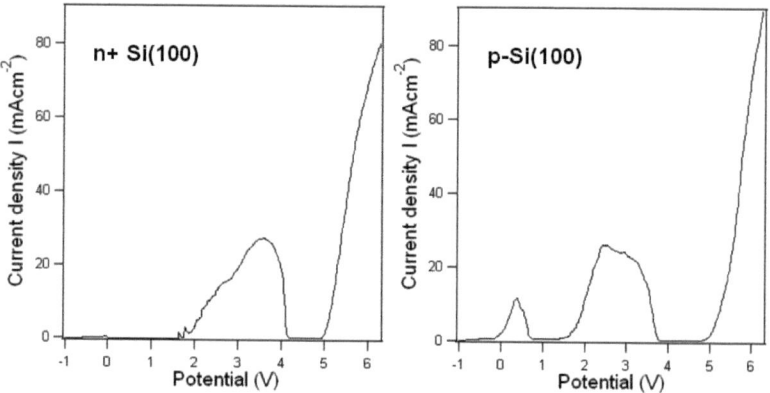

Fig. 3-71: Current-voltage behavior of (n+)-Si(100) and p-Si(100) in NH$_4$F (40%) in the dark (potentiodynamic scan with 20 mV/s). Two peaks are observable: behind OCP and near 3 V.

The relationship of crystal orientation to structure formation is summarized in Fig. 3-72. Here, optical micrographs are shown for n-Si(111), n-Si(100), n-Si(113) and p-Si(110). Except for p-Si(110), all preparations were carried out under illumination with about 7 mW/cm^2. The preparation time was 10 min each. As insets, the respective surface lattices are shown in approximate relation to the so-called primary flats, provided by the wafer manufacturer. While Si(100) surfaces almost always show square structures, independent on the initial surface etching and cleaning step, surfaces with more intricate surface lattice geometry as Si(111) and Si(113) tend to form randomly branching topographies. Si(111) samples with regular microstructures could only be produced after an optimized H-termination step as described in section 3.1.2. Si(113) samples, however, always exhibited random propagation of the branches, even after an improved electrochemical etching procedure carried out according to published results [221]. The surface lattice symmetry has therefore to be addressed as important parameter in model considerations to be discussed below.

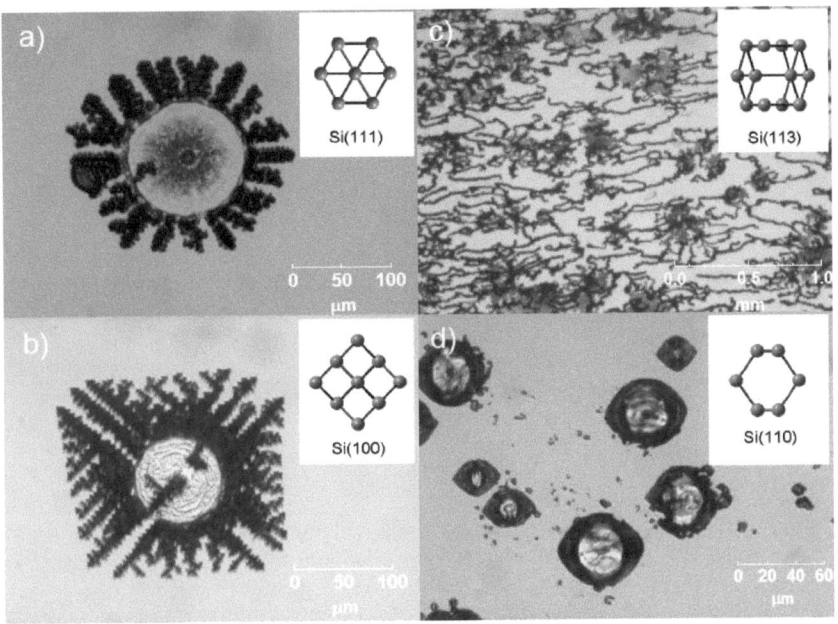

Fig. 3-72: Regular pattern formation obtained by potentiostatic treatment (6 V) for 10 min. (a) Hexagonal structure on n-Si(111). (b) Square structure on n-Si(100). (c) Lamellar structures on n-Si(113). (d) Rhombus-like structures on p-Si(110), prepared in the dark. The corresponding surface lattice geometries are shown as insets.

Variation of the NH_4F composition showed that structure formation requires concentrations higher than 20% (about 5 M). The current voltage behavior for varied concentrations, shown in Fig. 3-73, proves significant differences particularly in the region between the first current maximum and the current-plateau (U > 6 V) whereas the first current peak decreases proportional to the concentration. Although the current-plateau reaches similar values for all concentrations, no propagating of dendrites could be observed for 20% NH_4F.

Oversaturation to nominally 15 M induced microstructure growth on Si(111) as expected. The solution was stirred before starting the experiment to reduce the re-crystallization of NH_4F crystals at the liquid-ambient interface. The applied potential was 6 V and the conditioning time was 10 min.

Interestingly, the formation principle changed from hexagonal to triangular structure formation as shown in Fig. 3-74.

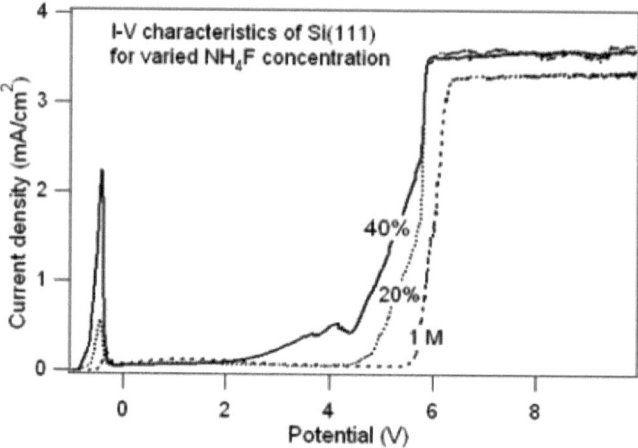

Fig. 3-73: Photocurrent-voltage behavior of Si(111) in varied NH$_4$F concentrations upon illumination with about 7 mW/cm^2. Scan velocity 10 mV/s.

Magnification of the corresponding HR-SEM image shows furthermore that the high regularity of the triangular shape is accompanied by an inner topography of porous material. The inset in Fig. 3-74 represents an area of about 15 µm length and shows holes with a diameter of 1 µm and below distributed across the interior of the triangular structures. Furthermore, the boundaries of single arms are markedly etched such that the regularity of the whole structure almost vanishes on this scale of magnification. Comparable results were obtained during potentiodynamic treatment of Si(111) samples. In this case, the branches started to overlap but still exhibited preferential growth along triangular directions. On Si(100) samples, however, no significant change of the structure formation could be observed. The topographies showed still regular squares with comparable density as obtained for lower concentrations of 40% NH$_4$F.

On back-side illuminated Si(111), hexagonal hole formation was observed as illustrated by the TM-AFM image in Fig. 3-75. The hole-structure comprises several substructures, concentrically aligned. Cross-sectional analysis of the maximum depth showed that the holes are extending more than 300 nm into the substrate. The piezo-crystal, moving the AFM tip, however, could not compensate these large distances and the actual depth might even exceed this value. The profile shown in Fig. 3-75b essentially determines the thickness of the anodic oxide layer built during the photoelectrochemical process. The oxide is characterized by recurrent nanopatterns of approximately 400 nm width. The arrangement looks like terrace alignment of triangular steps and demonstrates the pronounced inhomogeneity of the oxide growth.

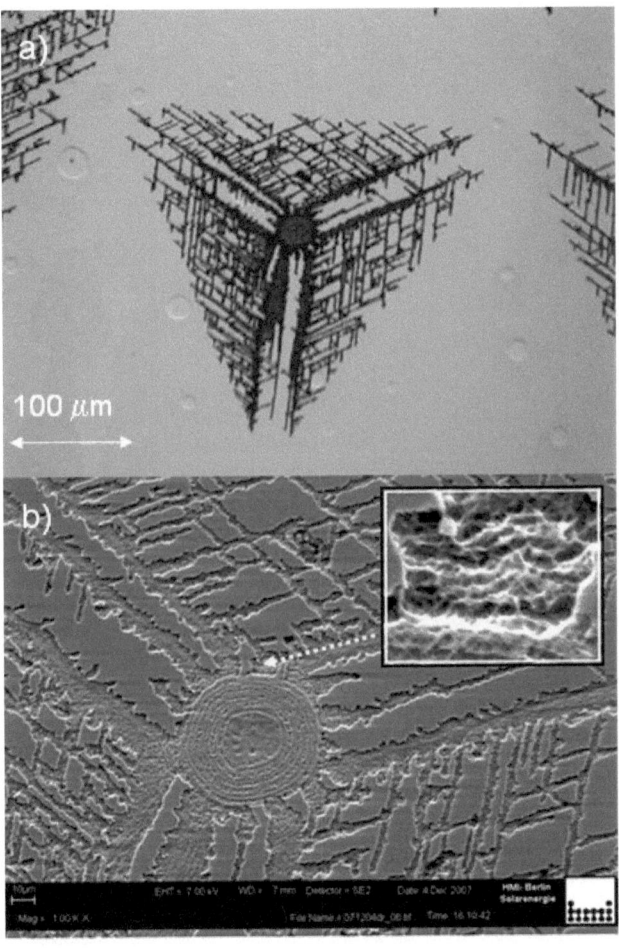

Fig. 3-74: Optical micrograph (a) and HR-SEM image (b) of microstructures on n-Si(111) obtained in oversaturated NH₄F solution. Nominal concentration: 15 M. The structures were observed after 10 min with illumination of about 7 mW/cm². The inset shows a magnification of approximately 15 μm length of the porous inner topography of the triangular structure.

In Fig. 3-76, a corresponding Si(100) structure, obtained after front-side illumination is shown. Usually, AFM images could not provide reliable height information for the microstructures since, as mentioned above, the tip could not follow the vertical variation across the surface. In this case, however, structure propagation appears suppressed;

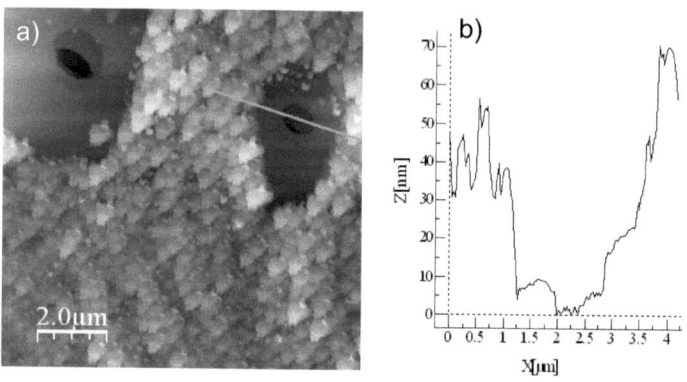

Fig. 3-75: TM-AFM image of a n-Si(111) sample after photoelectrochemical dissolution for 10 min using backside-illumination (left). The corresponding cross-sectional analysis (right) characterizes the height of the anodic oxide layer.

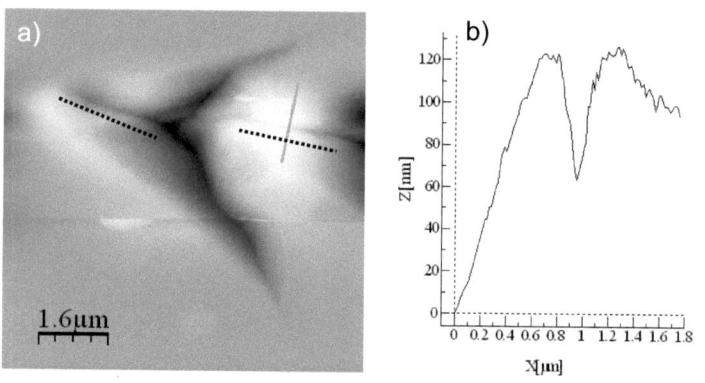

Fig. 3-76: TM-AFM image of a n-Si(100) sample after photoelectrochemical dissolution for 10 min using frontside-illumination (left). The corresponding cross-sectional analysis (right) shows nanocracks (above black dashed lines) propagating along the Si(100) crystal orientation.

the branches did not complete the lateral growth and single segments remained on the surface with smaller depth. The structures depicted here correspond to the tip of a Si(100) branch as shown, e.g. in Fig. 3-67a. Two segments are interconnected by a crack of about 50 nm depth as shown by the grey profile line in Fig. 3-76a and the cross-sectional analysis in Fig. 3-76b.

The two black dashed lines correspond approximately to the orientation of atomic bonds on the Si(100) surface lattice while the sidewalls of the microfacets show some deviation in their 45°-orientation relative to the propagation direction. These results will be of importance for the discussion of the structure formation principles further below.

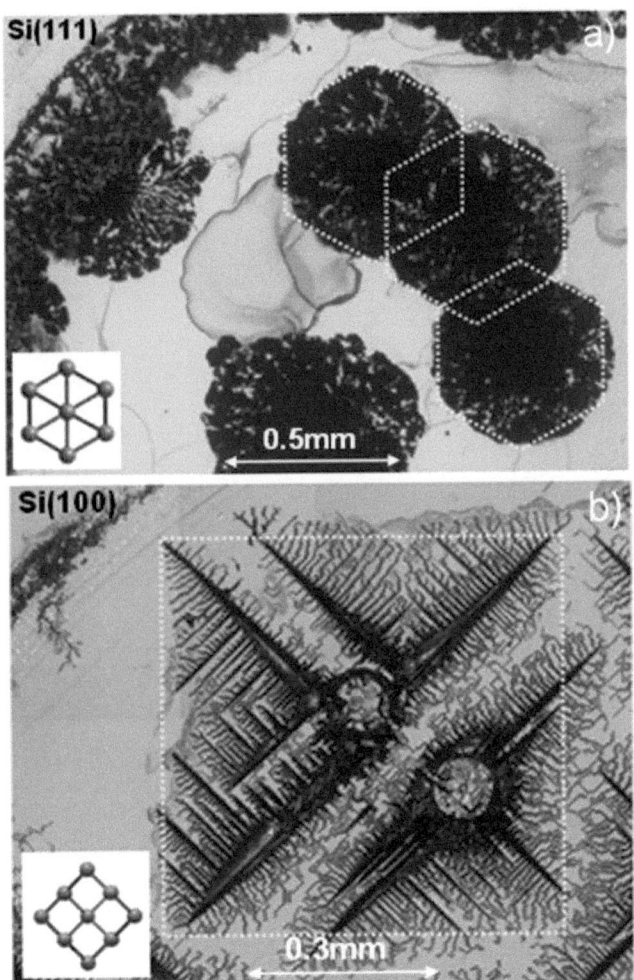

Fig. 3-76: Long-time conditioning of Si(111) (above) and Si(100) (below). The samples were imaged by optical microscopy after 60 min photoelectrochemical etching in 40% NH_4F. The resulting symmetries of the etch structures are emphasized by dashed lines forming hexagons and squares, respectively.

Extended photoelectrochemical conditioning times resulted in surface topographies as illustrated in Fig. 3-76. Despite increasing density of the branches (which appear to overlap in the case of Si(111)), the respective lattice symmetries are still recognizable. Dashed lines were added to the images in order to emphasize the hexagonal structure on Si(111) in Fig. 3-76a and, respectively, the square structure on Si(100) in Fig. 3-76b.

3.3.3 Influence of the photon flux on the surface topographic degree of order

Upon pronounced variation of the incident photon flux, the degree of regularity of evolving structures on Si(100), dramatically changes. This effect, observable at higher anodic potentials, is illustrated by Figs. 3-77a through d where U = 6 V was chosen.

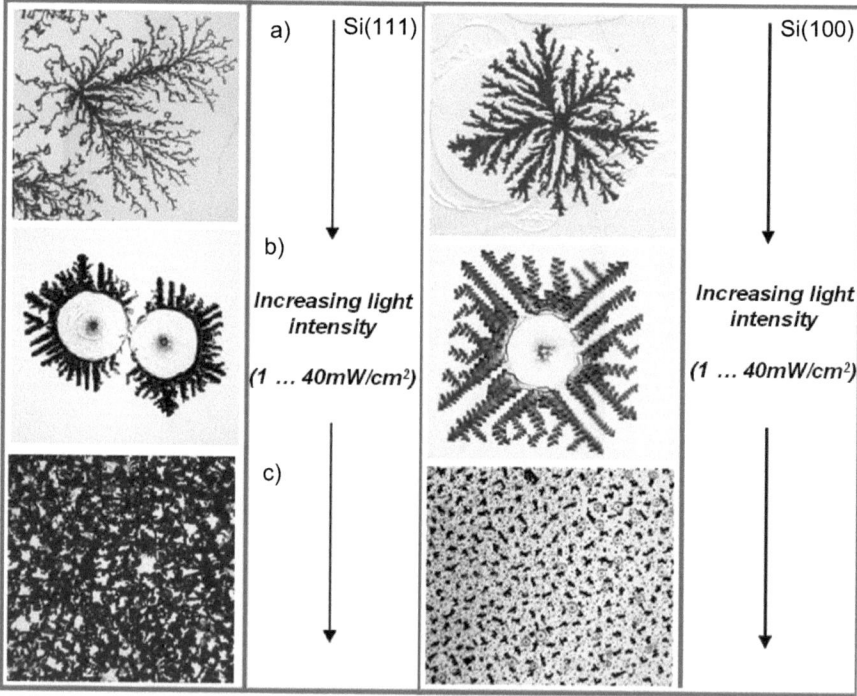

Fig. 3-77: Schematic representation of the transition in the pattern regularity upon incremented light intensities. Left: sequence of optical micrographs on n-Si(111). Right: corresponding images of Si(100) surfaces. Dendritic etch grooves are typical for low light intensities (first row of images) while more regular patterns require a medium photon flux (second row). Beyond a light intensity threshold, patterns start to overlap or to impede other patterns in the vicinity (third row left and, respectively, right).

Random branching is observable for low light intensities (below about 1 mWcm^{-2}) as shown in Fig. 3-77a. Si(111) surfaces were found to be more susceptible to this type of structure propagation than the Si(100) orientation. In a medium photon flux range of about 2 – 10 mWcm^{-2}, more regular structures are evolving (Fig. 3-77b). Again, regular structures on Si(111) surfaces were more difficult to prepare. For high illumination intensities (about 10 – 50 mWcm^{-2}), Si(111) surfaces exhibited strong corrosion while Si(100) surfaces were prepared with a higher density of small squares as well as partially completed branches. Overlapping of individual structures was not observed in contrast to Si(111) samples. The phenomenon of scaling, i.e., increasing structure number with decreasing structure size, was studied in more detail by light intensities which were successively incremented in small steps in the range of 2 to 8 mWcm^{-2}. The results for the Si(100) and Si(111) surface orientation are shown by assorted optical micrographs in Fig. 3-78.

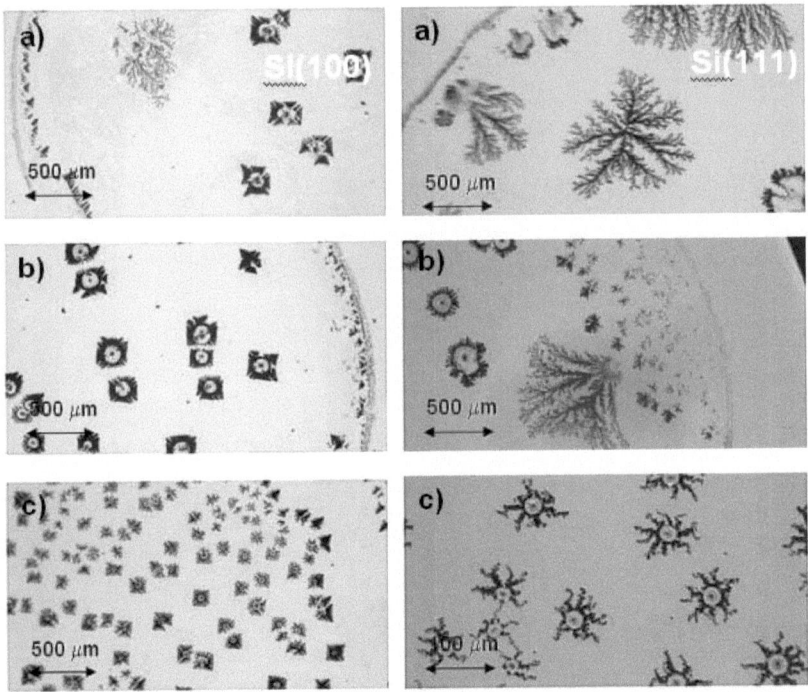

Fig. 3-78: Scaling effects on Si(100) and Si(111) for incremented light intensities of 2, 4 and 8 mW/cm^2. The resulting structure number increases from (a) to (c) with increasing light intensities. (c) The structures are beginning to reduce in size due to the dense structure distribution. The surfaces were all prepared in 40% NH$_4$F for 10 min at the potential U = 6 V.

Photocurrents were adjusted in the beginning of the experiments to the values indicated in the figures. The images in Fig. 3-78a, b and c exhibit increasing numbers of square etch structures which appear to mutually suppress their lateral growth as soon as the distance between adjacent structures falls below a certain value. Analysis of the corresponding chemical state at the surface of the Si(111) and Si(100) photoelectrodes was carried out by XPS measurements (see Figs. 3-79a through d).

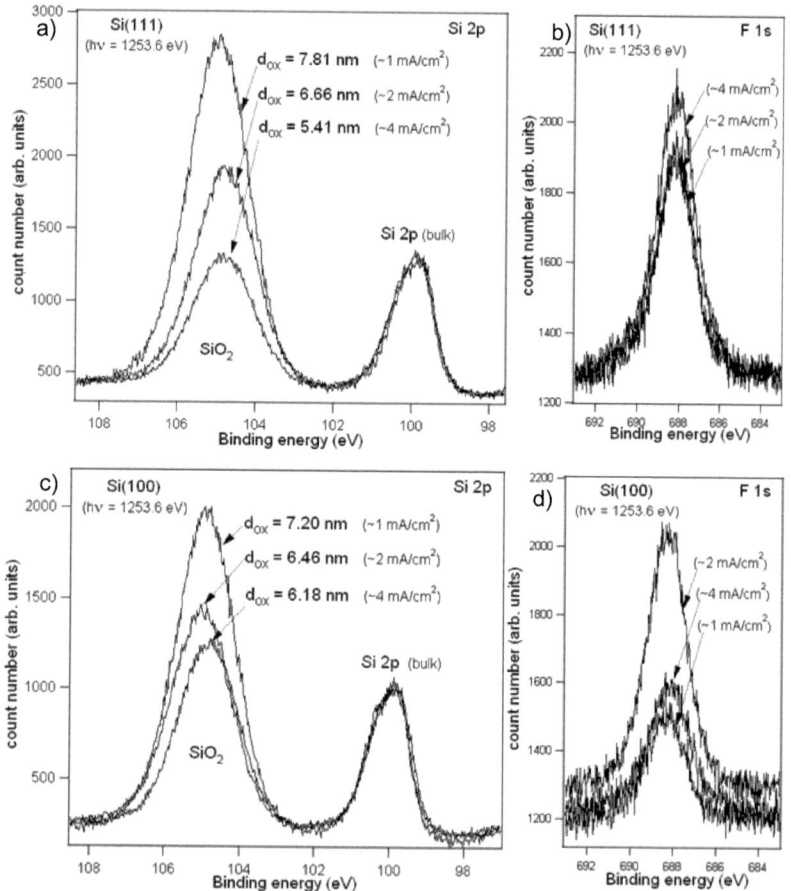

Fig. 3-79: XPS analysis of the Si 2p core level signal of the surfaces shown in Fig. 3-78 and the corresponding results for Si(111) samples. Additionally, measured F 1s signals are shown. The photocurrents, resulting from incremented light intensities, are indicated.

With increasing light intensities and, correspondingly, increasing numbers of structures, the ratio between the Si 2p core level signal ($E_B \sim 99.8$ eV) and the SiO_2 signal ($E_B \sim 104.5$ eV) decreases.

These results are providing a first indication that the structures may be covered by less silicon dioxide than surrounding areas. Therefore, a lower integrally measured SiO_2 signal is obtained for surfaces with more structures. Correspondingly, the analysis of the F 1s signal suggests the presence of, predominantly, Si-F_x compounds on the surface with a binding energy E_B near 688 eV [222]. While the sequence of F 1s signals for Si(111) suggests that the amount of Si-F_x increases with increasing structure number, i.e., Si-F_x is preferentially located within the structures, corresponding curves for Si(100) show a deviation of this relation for the case of 2 mWcm^{-2} illumination intensity (see Figs. 3-79b and d). Results from spatially resolved photoelectron spectroscopy using synchrotron radiation (PEEM) are shown in Figs. 3-80a through c.

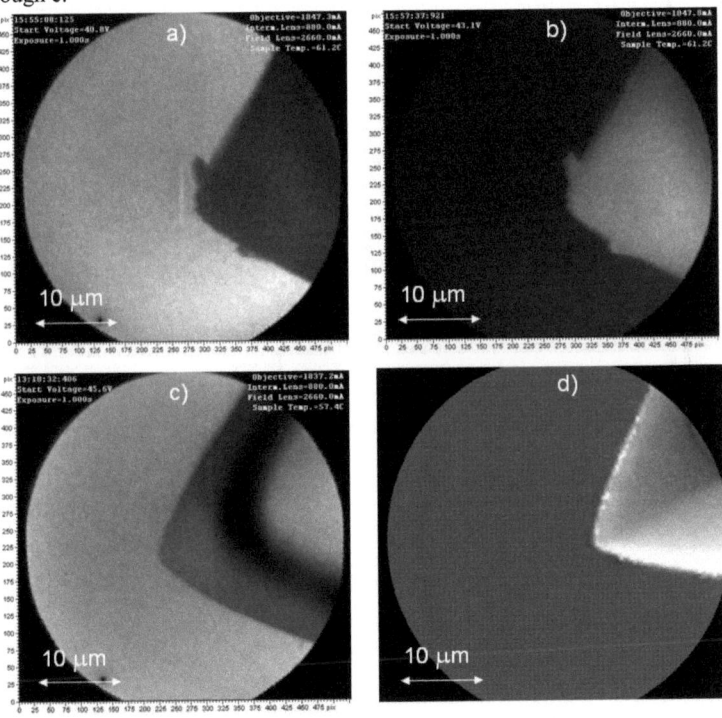

Fig. 3-80: PEEM analysis of the spatial distribution of SiO_2 and Si by measurement of the Si 2p core level signal. (a) The photoelectron yield and therefore the contrast is determined by the presence of SiO_2. (b) A reversed contrast is obtained by detection of the Si bulk signal. (c) The surface was investigated after an HF etching for 2 min which removed the SiO_2 layer. (d) HR-SEM image of the surface before etching.

The analyzed area corresponds to a tip of a Si(100) structure as shown in Fig. 3-67a. The contrast in these images is provided by the photoelectron yield of emitted electrons within a small kinetic energy range. In Fig. 3-80a, the detection energy was chosen in order to measure the yield of electrons with a binding energy of about 104 eV upon excitation with hv = 150 eV. The image reflects therefore the distribution of SiO_2 across the surface. The photoelectron yield inside the structure is distinctively lower than outside the structure. By selection of a lower detection energy, corresponding to the silicon bulk signal of the Si 2p core level at $E_B \sim$ 99 eV, the contrast of the image appears reversed. Finally, after exposure of the surface to hydrofluoric acid (50% HF for about 2 min), the contrast almost vanishes except for the boundary between the tip and the surrounding area (see Fig. 3-80c). The remnant area in darker grey is most probably due to the alignment of the X-ray source with respect to the surface of the sample with a small inclination angle.

3.3.4 On the role of interfacial stress on fractal structure propagation

The formation of the etch structures shown, e.g., in Figs. 3-64 and 3-66 is obviously caused by elevated etch rates k_z, perpendicular to the surface, and $k_{x,y}$, parallel to the surface. Profilometry results which were presented in the previous section provide depth information across the interior of the structures. According to these measurements, the structures extend down to 6 μm on Si(100) samples. Using HR-SEM information about the lateral extent of the structures (see Figs. 3-64 and 3-66), etch rates of $k_z \sim 10$ nms^{-1} and $k_{x,y} \sim 300$ nms^{-1} can approximately be derived. For Si(111) samples, these results presumably have the same order of magnitude. Compared to the chemical etch rate of Si(111) in air saturated 40% NH_4F at room temperature, the dissolution upon photoelectrochemical conditioning is increased by a factor of 10^3 for k_z and 3 x 10^4 for $k_{x,y}$. The influence of oxygen dissolved in the NH_4F solution, on the other hand, is reported to increase the chemical etch rate for the Si substrate but to decrease the corresponding etch rate for SiO_2 (by a factor of about 10 each). The evolution of oxygen during the photoelectrochemical dissolution may therefore play an important role for the structure formation by increased oxidation rates and varied etching rates. Estimating the number of atomic bonds that have to be dissolved in order to produce a Si(100) structure as shown in Fig. 3-66, a total charge of the order of only 1mC is required. The oxide overlayer thickness of below 10 nm, as determined by photoelectron spectroscopy, in combination with the oxide etch rates of anodic oxides in 40% NH_4F (see chapter 3.2), shows that SiO_2 formation is equally associated with a charge flow of only a few mC.

Therefore, oxygen formation represents a considerable contribution to the total minority charge carrier consumption in experiments as described by Fig. 3-63b where the total charge flow amounts to about 1C.

Integrally measured XPS signals on surfaces with increasing structure number (see Figs. 3-78 and 3-79) suggest that the structures locally remove the anodic oxide layer. Therefore, it is tentatively assumed that the structure propagation occurs in response to a force that originates in the oxide layer. It is well known that there exists a volume mismatch of about 2 between the SiO_2 layer and the underlying silicon substrate. Moreover, much work has been carried out to detail the resulting stress and strain energies at the SiO_2/Si interface [5, 41, 43, 160]. Although most results are related to thermal oxides, also anodic oxides can be assumed to invoke, at a lower magnitude, stress effects at the interface as discussed before.

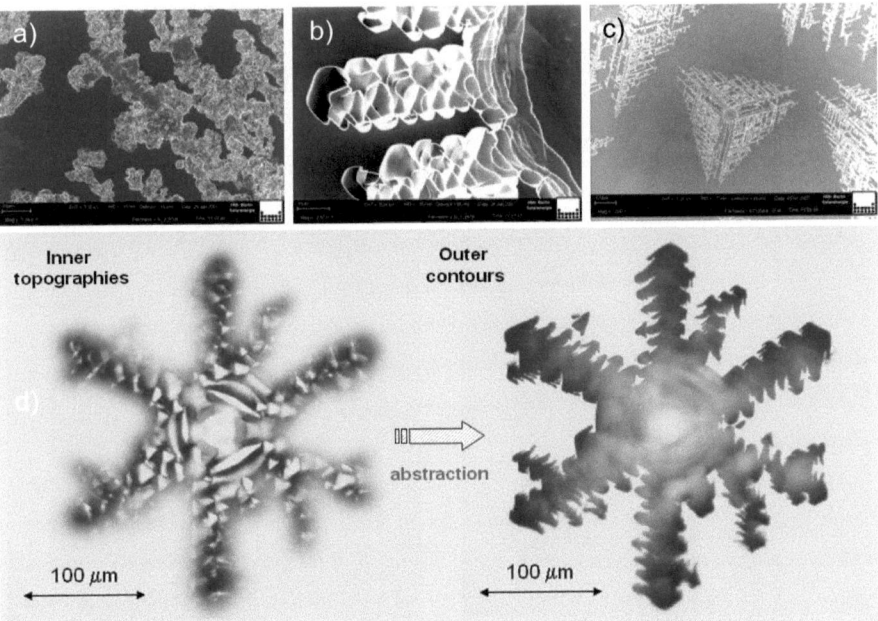

Fig. 3-81: Scheme for the model development of fractal microstructure formation. The assorted HR-SEM images above show varying combinations of (ir)regular outer and (ir)regular inner topographies prepared by different experimental conditions. The process of abstraction, illustrated by the lower image, is based on the assumption that, in a first approach, the formation of outer contours can be analyzed and modeled without consideration of the inner topographies.

In order to investigate the highly anisotropic dissolution behavior, proven by HR-SEM and optical microscopy, reasoning is sought for the interdependence between the etch rates k_z, $k_{x,y}$, on one hand, and the properties of the respective silicon lattice, on the other hand. This interdependence has already been shown by the respective two-fold and three-fold symmetry of the etch structures observable in Figs. 3-67a and b.

If the oxide removal can be regarded, in terms of dynamic system theory, as a response to a force-like quantity originating in the oxide layer, then a feedback-like interaction is identified that may be responsible for the observed self-organization phenomena during structure formation. According to Fig. 3-81, fractal structures proved so far independent formation principles of regular inner and outer topographies. Image 3-81a shows randomly branching structures on Si(111) characterized by smoothly polished microfacets. Image 3-81b illustrates regular contours enclosing regularly etched segments on Si(111). From experiments with oversaturated solutions it is known, on the other hand, that regular triangles on Si(111) preserve their geometric arrangement despite pronounced porosity in the interior. It might therefore reasonable to base initial theoretical considerations on the propagation process of the branch contours only.

This approach is illustrated by Fig. 3-81d where varying focal planes of the optical microscope were used to emphasize the complex inner structure and the pattern of the outer contours respectively. The process of abstraction in the subsequent model follows the assumption that the interior of the structures can be ignored.

It is assumed that atoms at the edge or tip of a fractal structure endure mainly lateral strain or distortion since they are not fully coordinated by other Si atoms. This assumption is illustrated by Fig. 3-82, left, where a cross section of an area close to a fractal tip is schematically depicted.

The volume mismatch between the oxide layer and the bulk material in the vicinity results in lateral strain and distortion of the topmost atomic bonds. Displacement of the atoms and the direction of crack propagation are indicated in the figure. According to theoretical approaches for crack propagation in brittle material, published during the 1980's and 1990's [223-225], strain across a surface-near region is assumed to facilitate the propagation of the cracks. In the case discussed here, nucleophilic attack of strained bonds in the presence of photogenerated holes and water is thought to be more probable during the photoelectrochemical dissolution process.

The HR-SEM image of a fractal Si(100) tip in Fig. 3-82, right, shows schematically the lateral strain that points in opposite direction of the structure propagation: balls are indicating atom positions while arrows point in the direction of the resulting strain.

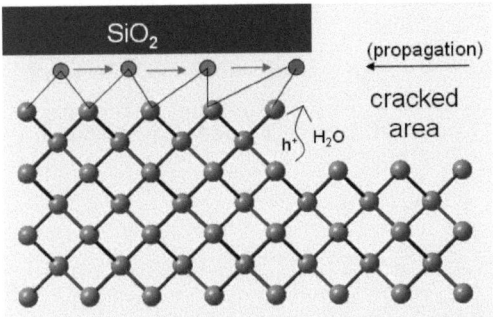
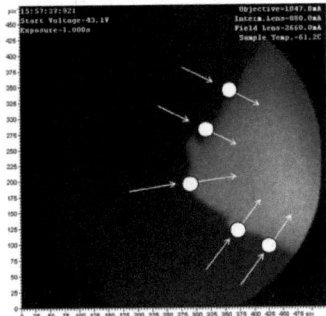

Fig. 3-82: Schematic cross section (left) of a region near a fractal tip. Strain of the atomic bonds is assumed to facilitate nucleophilic attack of the silicon atoms in the presence of photogenerated holes and water. Right: PEEM image of the fractal tip with arrows indicating the displacement of atoms (white balls) due to SiO_2 induced strain.

It should be noted that this effect has to be distinguished from the stress distribution at the SiO_2/Si interface in diluted NH_4F solutions where photocurrent oscillations are observed: in the latter stress and strain are acting less preferentially and also vertical stress components at the SiO_2/Si interface presumably contribute to the oscillating layer growth of porous SiO_2.

3.3.5 Numerical simulation of the microstructure formation

In the following, the possible effect of oxide induced interface stress and strain on the structure propagation will be analyzed by computations on a two-dimensional square grid in analogy to the Si(100) surface orientation. Normal and shear stresses directed perpendicular to the x-y plane are assumed to be zero (plane stress). Model considerations are furthermore based on the assumption that the stress-strain relation for isotropic elastic media holds, i.e., the equation

$$\sigma = D\varepsilon \qquad (3\text{-}14)$$

describes correctly the physical conditions. Here, σ denotes the stress tensor

$$\sigma = \begin{pmatrix} \sigma_x & \sigma_{xy} \\ \sigma_{xy} & \sigma_y \end{pmatrix} \qquad (3\text{–}15)$$

and ε the corresponding strain tensor:

$$\varepsilon = \begin{pmatrix} \varepsilon_x & \varepsilon_{xy} \\ \varepsilon_{xy} & \varepsilon_y \end{pmatrix}. \qquad (3\text{–}16)$$

Eq. 3-14 expresses the stress-strain relationship, i.e., elongations of the form $\varepsilon_x = \dfrac{\partial u}{\partial x}$, $\varepsilon_y = \dfrac{\partial v}{\partial y}$ and $\varepsilon_{xy} = \dfrac{\partial u}{\partial y} + \dfrac{\partial v}{\partial x}$ are proportional to the stress exerted onto the solid. The tensor D depends on the material properties, i.e. the modulus of elasticity, E, and Poisson's ratio, v:

$$D = \frac{E}{1-v^2} \begin{pmatrix} 1 & v & 0 \\ v & 1 & 0 \\ 0 & 0 & \frac{1-v}{2} \end{pmatrix}. \qquad (3\text{-}17)$$

The set of differential equations that has simultaneously to be solved in order to calculate the equilibrium positions of surface atoms after initial displacement is defined by:

$$G\left(\frac{\partial^2 u}{\partial x^2} + \frac{\partial^2 u}{\partial y^2}\right) + G\frac{1+v}{1-v}\frac{\partial}{\partial x}\left(\frac{\partial u}{\partial x} + \frac{\partial v}{\partial y}\right) = 0 \text{ and} \qquad (3\text{-}18)$$

$$G\left(\frac{\partial^2 u}{\partial x^2} + \frac{\partial^2 u}{\partial y^2}\right) + G\frac{1+v}{1-v}\frac{\partial}{\partial y}\left(\frac{\partial u}{\partial x} + \frac{\partial v}{\partial y}\right) = 0 \qquad (3\text{-}19)$$

with $G = \dfrac{E}{2(1+v)}$.

Following this qualitative approach, one of the fundamental effects associated with the structure formation can be quantitatively assessed: in Fig. 3-83, two approaching structures on a Si(100) sample are shown which are characterized by two phenomena. Firstly, the branches which are extending from the center point are continuously ramifying but they do not overlap. Secondly, when approaching each other, the structures decelerate the ramification process towards each other. The latter phenomenon is observable for structures, too, which show more random propagation as depicted in Fig. 3-77a. Subsequently, strain analysis was carried out, simulating two adjacent square structures whose boundaries are affected by initial strain pointing inwards the squares. These initial conditions are indicated in Fig. 3-83b by arrows. The following calculation determines distortions of the elements of a square grid, u(x,y) and v(x,y), in x- and y-direction only (in-plane stress analysis). Initial (and equal) displacements at the boundaries of the two square structures are assumed. The calculation results in a relaxation of the grid with minimized stress and strain energies. In Fig. 3-83b, the resulting total displacements, i.e. $u^2 + v^2$, are indicated by a corresponding grey scale. Iso-displacement curves are added in darker grey. It can be seen that deformation of the grid is lowest between the squares and highest at the corners. This analysis shows, if oxide induced stress is assumed, relaxation of the stress and strain energies is most effective in an area where the structures approach each other. This result is true for the same reason for individual branches of a single structure and may account for the fact that branches generally do not overlap. The surface lattice of Si(111), however, is much more complicated than the cubic Si(100) lattice. Perturbations of the regular structure propagation in this case are more probable and this explains why randomly ramifying structures are more often observed on this surface orientation.

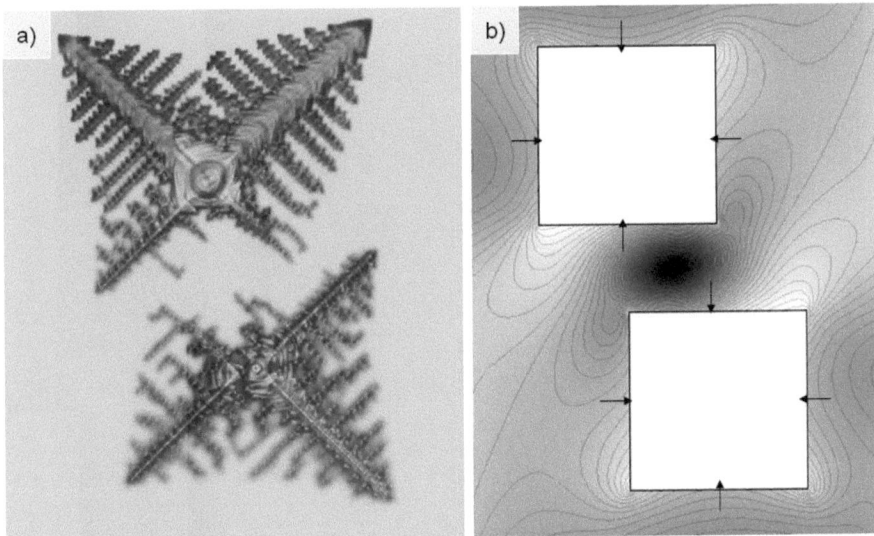

Fig. 3-83: (a) Two approaching fractal structures on Si(100). (b) In-plane stress analysis for two square holes embedded in a silicon slab. The grey scale refers to the resulting total displacements after lattice relaxation. Iso-displacement curves are indicated by darker grey (see discussion). Initial conditions are indicated by arrows pointing in the direction of the assumed displacement.

The calculation, illustrated by Fig. 3-83b, represents the time-independent case only. Time-dependent simulations would have to consider the removal of atoms from the grid which show the highest displacement (and therefore higher dissolution probability). Rather than performing the complicated numerical calculation in successive steps, the results of the in-plane stress analysis is used in the following to design a simple but suitable simulation procedure which allows prediction of the structure formation for a given surface lattice.

If a homogeneous oxide layer is assumed, covering the surface around the structures, the induced stress may also be considered as uniform, i.e., isotropic.

The propagation of the branches may then occur at a constant velocity v_x and v_y. Oxygen bubbles, adsorbed to the surface in the very beginning of the experiment, are thought to provide the initial cracks as discussed before. The structure growth can continue as long as the distance between two or more branches do not fall below a given value, otherwise a branch has to change direction or to stop at all in its growth. A simulation based on these assumptions is shown in Fig. 3-84b. Comparison to a structure built at the potential $U = 6$ V in Fig. 3-84a shows that both the distribution of branches as well as the relative orientation to the Si(100) lattice are almost perfectly achieved.

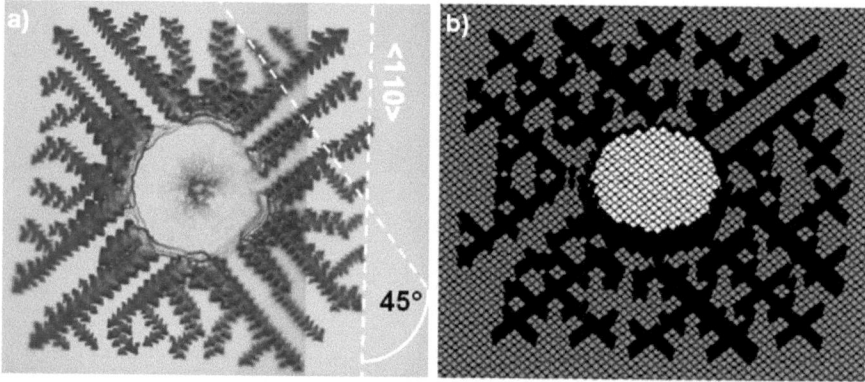

Fig. 3-84: Comparison of an experimental structure obtained in 40% NH$_4$F after 10 min at $U = 6$ V (light intensity 7 mWcm^{-2}) with a simulated structure. (a) The orientation of the fractal structure with respect to the surface lattice is indicated by an angle which refers to the primary flat of the wafer. (b) Simulated structure assuming uniform propagation of the branches which extend from initially cracked sites located at the boundary of an adsorbed oxygen bubble.

The orientation of the structure is referred to the <110>-direction which is usually identifiable on the wafer as so-called primary flat. The rotation angle of 45° with respect to the atomic bond orientation, in turn, results from the uniform propagation of the branches at a constant velocity. The same principle that hinders individual branches of single fractal structures to overlap accounts for another observed phenomenon: if two ore more structures approach each other, the growth rate decelerates if a certain mutual distance is reached. It can be seen in Fig. 3-85b and c that for both regular and random propagation the overlap of branches is suppressed.

This formation principle is the same as for the individual branch growth within single structures. The corresponding simulation therefore has not to add any further assumption in order to predict the growth of two approaching structures. The computer based simulation was carried out for two center areas closely arranged to each other. The limit for the mutual distance suppresses crossing of the branches in agreement with experimental observations.

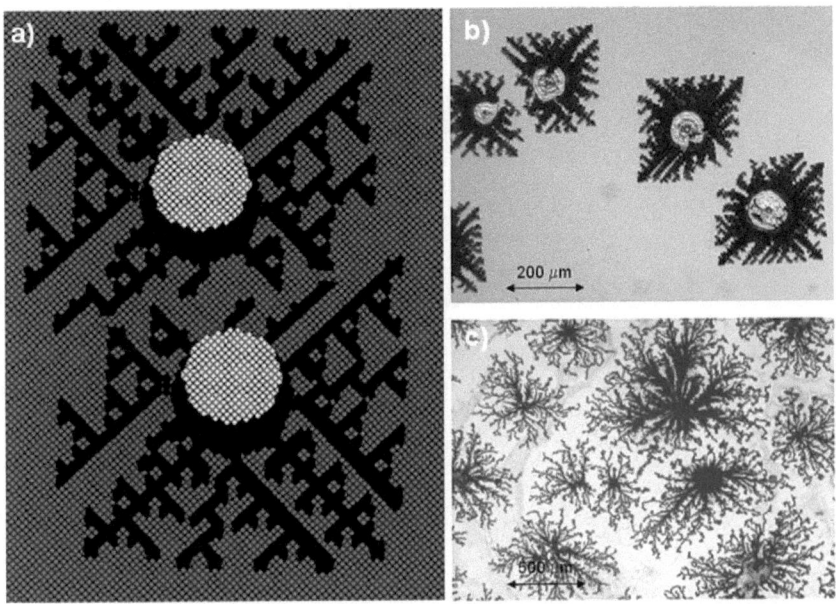

Fig. 3-85: Effect of approaching Si(100) structures. (a) Simulated distribution of branches. (b) Experimental results for regular structure growth on Si(100). (c) Experimental results for random structure growth on Si(111). Overlapping of individual branches is suppressed in (b) and (c). This effect can be simulated by the model described in the text.

Inhomogeneous oxide coverage was finally assumed for the simulation shown in Fig. 3-86. The inhomogeneity can be attributed to the lower minority charge carrier supply upon low photon flux conditions corresponding to Fig. 3-77a where the oxidation rate may be too low to produce a homogeneous layer. In fact, Fig. 3-86a, as a magnification of Fig. 3-77a, shows some inhomogeneous structures around the ramifying branches. Currently, it is not known if low light intensities result, on average, in less thick oxide layers. The influence of oxygen and H^+ generation on the (pH-dependent) oxide etch rate may play a crucial role for the integral oxide thickness. More important, however, is the local variation of the thickness with time. In this case v_x and v_y become time-dependent, too, and they mostly do not coincide, i.e., $v_x(t) \neq v_y(t)$ for almost all the time.

This effect is realized in the numerical simulation by an additional random function which varies v_x and v_y within a predefined range. As a result, a structure as depicted in Fig. 3-

86b is obtained. Again, the agreement with the experimental finding shown in Fig. 3-86a is very good, uncovering the possible origin of the random structure growth.

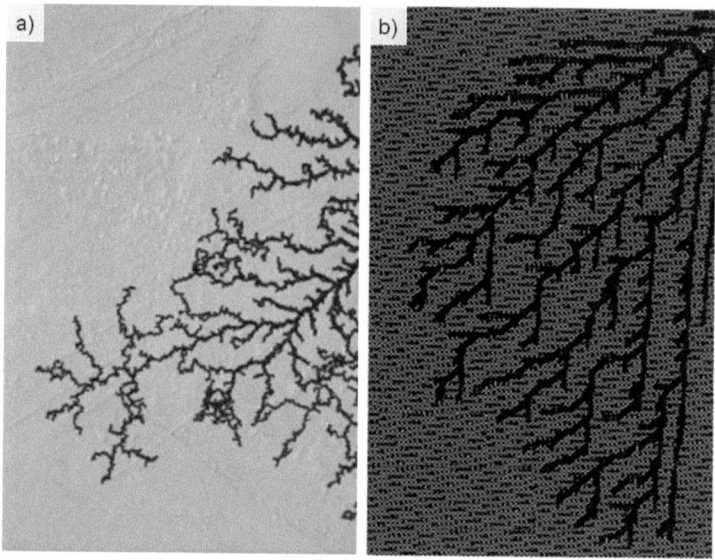

Fig. 3-86: Comparison of an experimental and simulated structure for the case of lower light intensity (below 1 mWcm^{-2}). According to model considerations, the lateral propagation velocities of the branches are varying with time resulting thus in less regular structures.

3.3.6 Summary and comparison of lateral with vertical macropore formation

The formation of fractal-like microstructures, prepared by photoelectrochemical preparation in concentrated NH$_4$F, was investigated for varied experimental conditions. Despite the intricate inner and outer geometries of the structures, propagation of the outer contours could be modeled with high agreement by a stress related model. This model is based on photoelectron analyses of the oxide distribution in integral and spatially resolved measurements. The finding of less thick surface oxides within the microstrucures suggested stress induced atom displacement at the edges of the structures and hence an increased dissolution rate. Numerical computations of the stress field accounted for most of the experimental findings, i.e. scaling effects and stress relaxation between adjacent structures and, respectively, between individual branches. The simulation was carried out for the Si(100) surface orientation but can seamlessly be transferred to more complex surface lattice geometries as the Si(111) orientation. The findings complement observations of vertical macropore formation insofar that fundamental formation principles as branching, relation to the crystal structure, inhibition of overlapping of individual branches and pore size have their respective equivalence.

However, does the stress model, presented here, also adequately describe nucleation and growth of the macropores? This question arises since stress forces and atom displacement (strain) is a phenomenon already proven for porous silicon [186-188]. Particularly, tensile stress during macropore propagation is a constituent in the model of Parkhutik: vertical cracks are forming at a macropore's tip, inducing tensile strain in the adjacent bulk material; hydrogen migration towards these bulk vacancies leads to silicon hydride formation and enhanced dissolution [226, 227]. This model is illustrated by Fig. 3-87.

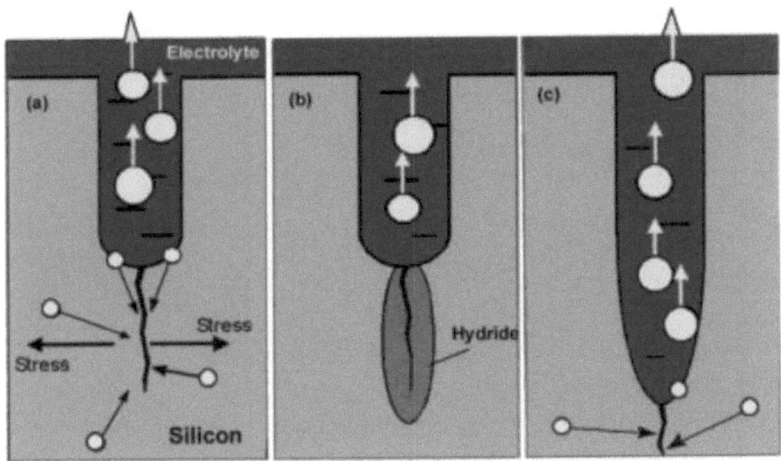

Fig. 3-87: Model of stress assisted macropore propagation proposed by V. Parkhutik [227]. (a) Migration of hydrogen atoms, either from silicon bulk or the electrolyte, to a stress induced microcrack. (b) Formation of silicon hydride in the cracked region. (c) Pore propagation towards the cracked region due the increased chemical activity of silicon hydride with respect to HF.

In the *current-burst model* a more complex interplay of the chemical state of the outmost silicon atoms and the shape of the space chare region in the vicinity of the pore tips is considered. Spatially and temporally inhomogeneous current flow is proposed leading to either direct silicon dissolution or oxidation with immediate dissolution of the oxide [86]. H-termination results in surface passivation and acts - according to the model - as synchronization force. Without H-termination, there are many surface states and the Fermi level is pinned. As a consequence, no pronounced space charge region is present as required for focusing the charge carriers to the pore tip. The situation changes with completed H-termination of the atoms at the pore walls. Interestingly, the authors of this model also observed the formation of so-called etch domains [228]. Regular domains were found on GaAs samples while silicon exhibited random branching only. However, the macropores, arranged in these domains, are still well-separated, i.e. the side-walls of the structures were intact in contrast to the uniform spreading of the lateral structures discussed here.

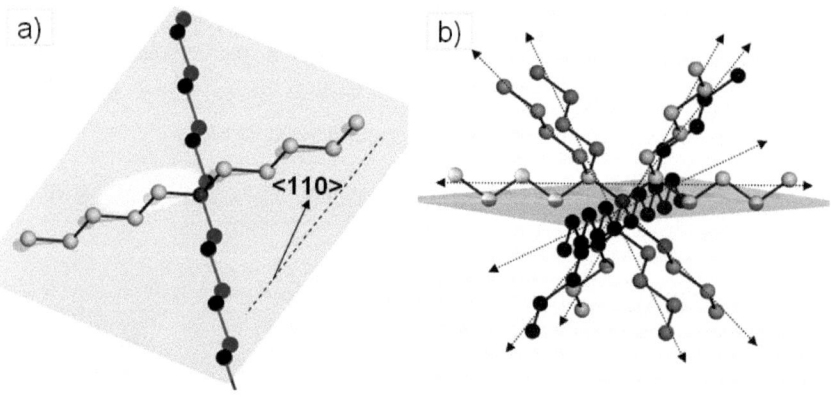

Fig. 3-88: Configuration of *atomic chains* on a Si(100) surface. (a) In-plane chains depicted with respect to the <110>-direction (perpendicular to the primary flat on a Si(100) wafer disk). (b) Indication of four further atomic chains located within the (111), (-111), (1-11) and (11-1) planes.

In Fig. 3-88a, selected surface atoms of the Si(100) surface orientation are shown. Only atoms were chosen having bonds along the surface plane. These atoms form linear *in-plane chains* which fill, by translational symmetry, the whole surface area. The orientation of the chains is given with respect to the <110>-direction.

According to the results of sections 3.3.2 through 3.3.5, stress energies are transferred along these atomic in-plane chains and the structure growth follows their orientation. Other chains, as shown in Fig. 3-88b, were ignored according to the model abstraction (Fig. 3-81) which considered outer contours only. Comparison with a model-macropore in Fig. 3-89 reveals a considerable difference between propagation direction and the underlying atomic configuration. Fig. 3-89a shows a scheme of a macropore which propagates along the <100>-direction. The image is compared to macropores obtained by experiments on p-type (b) and n-type (c) silicon, respectively. The atomic chains, illustrated by Fig. 3-88, are inclined in this case with respect to the propagation direction and the plane of the depicted cross-section.

In comparison to the lateral formation of microstructures, this observation points to an important difference of the propagation process. While the direction of lateral structures coincides with the direction of atomic bonds at the surface, vertical pores are inclined with respect to the bond configuration.

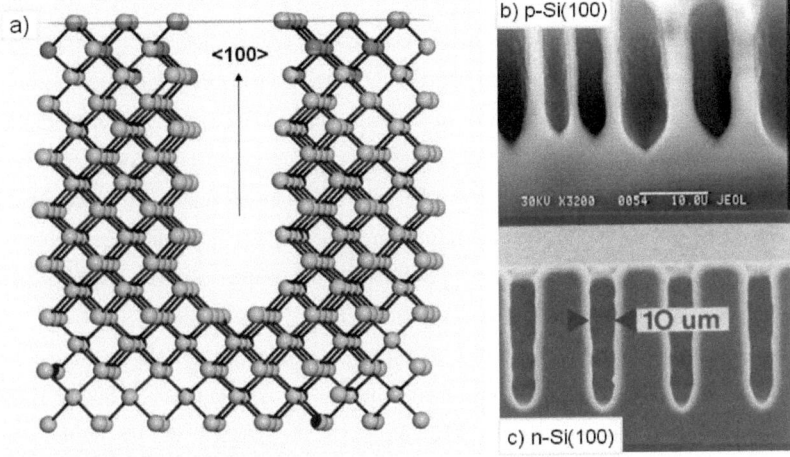

Fig. 3-89: Vertical macropores on a Si(100) substrate. (a) Scheme of the atomic structure of a Si(100) macropore. (b) and (c) SEM images of Si(100) macropores observed for p- and n-type Si(100) [229].

Although stress may contribute, to some extent, also to the formation of macropores, further assumptions appear to be inevitable to model the corresponding formation principles.

Future investigations have to clarify if these additional assumptions can also be related to the principles which determine the evolution of inner topographies of the fractal structures which were not considered in this work.

Summary

In the presented work, detection, assessment and modeling of stress related phenomena at the silicon dioxide / silicon interface in the presence of fluoride containing solutions was discussed. In section 3.1, Brewster-angle analysis was introduced as a surface/interface sensitive technique and compared to results from synchrotron radiation photoelectron spectroscopy. The proven sensitivity in the sub-angstrom region allowed for monitoring of the accelerated dissolution of the SiO_2/Si interface in ammonium fluoride containing solutions. It could be shown that the increased dissolution was independent of the used concentration and specific to the compactness of the investigated oxide. Therefore, this phenomenon could be related to lattice deformations in a sub-surface region which are reportedly induced by oxide induced stress forces. Based on these results an enhanced wet chemical preparation method could be developed by which the stepped Si(111) topography showed optimized parallel terraces while the chemical surface state was characterized by less suboxides.

The real-time BAR monitoring technique was in the following employed for the assessment of the photoelectrochemical dissolution of n-type silicon electrodes in diluted solutions of ammonium fluoride. The observed dependence of the current-voltage characteristic upon increased light intensities was exploited to develop a new surface conditioning technique. Nanotopographies, built in the divalent dissolution region, were subsequently oxidized in order to manipulate structure density and shape. It could be shown that the structures were embedded within an oxide layer. The following removal of the layer exposed the manipulated nanostructure fields. As one main result, a relation could be proven between the miscut angle of the used Si(111) wafers and the resulting distribution of the nanostructures. Miscut angles towards the $<11\bar{2}>$ direction resulted in single nanodots aligned in correspondence to the original orientation of the terraces. Accordingly, large miscut angle variation towards the <011> direction resulted in curved rows of nanostructures in analogy to the stepped topography of the used wafer. Finally, a qualitative model was introduced in order to explain the increase in the aspect ratio of single nanostructures. According to the findings of the preceding section, stress forces induced by local oxidation were proposed to result in preferential oxidation of the nanostructure side-walls.

The dissolution of silicon photoelectrodes in the tetravalent region was investigated with respect of topography formation at the electrolyte/oxide and the oxide/substrate interface. Again, BAR was employed to correlate the oscillation cycles of the photocurrent and the thickness of the anodic oxide layer. The evaluation of the reflectance data was facilitated by a mathematical approach which relates surface transformations, such as oxide, roughness or porous layer formation to the corresponding reflectance by consideration of the charge flow. The onset of the reflectance behavior was interpreted as instantiation for the initial formation of a stressed interfacial region. This region was exposed by subsequent etching steps and was markedly characterized by recurrent microstructures.

The novel finding of lateral microstructures produced by anodization in concentrated ammonium fluoride solutions was finally investigated. By variation of the substrate orientation, the distinct dependence of the structure symmetry on the surface lattice configuration could be proven. Transitions in the forming principles, from random branching over regular propagation to size-limitation upon light intensity variation, could be related to integrally measured oxide thicknesses. It could be shown that less oxide was measured on electrodes with higher structure density. The conclusion that interior structure areas are covered by less oxide could be proven by photoelectron spectroscopy with spatial resolution. The impression that the propagation process "intentionally" removes oxide from the electrode area was physically interpreted as response to the presence of a stress field. In model calculations, the principal properties of the stress field were determined. It could be shown that strain is concentrated at the corners of square shaped model structures. Furthermore, stress and strain attenuation was calculated for approaching individual structures in correspondence to experimental observations. Stress release was thereby identified as possible cause of the scaling effects upon high incremented light intensities. According to these results, a simplified computer model was developed in order to simulate the structure propagation. With the additional assumption that the homogeneity of the anodic oxide layer influences the feedback mechanism, all major experimental findings could be explained, i.e. the development of self-similar branches, the orientation with respect to the surface lattice, scaling effects and the statistical branch distribution upon low light intensities.

Appendices

A.1 Multi-layer analysis of Brewster-angle data

Instrumental imperfections of the BAR setup disturb the perfect parallel alignment of the light beam. This divergence leads to, firstly, a distribution of the angle of incidence and, secondly, a distribution of the state of polarization. In model experiments, the angle distribution with respect to the optical axis was determined. Diaphragms with decreasing diameter from 1 to 0.2 mm were illuminated by the probing light and the angle dependent reflectance behind the diaphragms was measured [230]. Data evaluation suggested a square-like distribution profile, $Z(x)$, with angle divergence of $\alpha \approx \pm 5$ mrad as shown in Fig. A1-1. $Z(x)$ remains constant over the width $2w$ and is characterized by an almost linear slope in the interval of width Δ. In Fig. A1-1 the measured intensity distribution is compared to the results of a simulation, based on a convolution of $Z(x)$. Except for scattering effects, good agreement could be achieved.

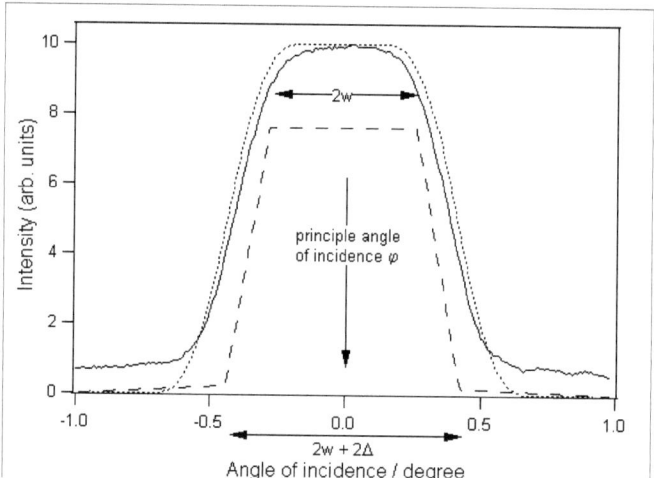

Fig. A1-1: Determination of the angle divergence of the probing light beam. Solid line: measured intensity distribution behind a diaphragm of 0.5 mm diameter. Dashed line: calculated intensity distribution. Dotted line: simulated intensity by evaluation of a convolution integral. Except for scattering effects for larger angles, good agreement between measured and simulated data could be achieved.

Measured BAR data represent therefore a convolution of individual reflectance values around the principle angle of incidence. Correspondingly, the state of polarization varies around the p-state with nearly identical distribution, $Z(\varphi)$.

An incident electromagnetic plane wave at arbitrary polarization angle can be described by the *Jones Formalism* [91] as:

$$\vec{E}_i = \begin{pmatrix} \cos x & -\sin x \\ \sin x & \cos x \end{pmatrix} E_0 \begin{pmatrix} 1 \\ 0 \end{pmatrix} = E_0 \begin{pmatrix} \cos x \\ \sin x \end{pmatrix}. \quad (A1\text{-}1)$$

Eq. A1-1 describes an electromagnetic wave initially p-polarized and rotated by the angle x. After reflection at a surface the Jones vector can be expressed by the respective complex reflectance coefficients of the material:

$$\vec{E}_r = \begin{pmatrix} r_p(\varphi) & 0 \\ 0 & r_s(\varphi) \end{pmatrix} E_0 \begin{pmatrix} \cos x \\ \sin x \end{pmatrix} = E_0 \begin{pmatrix} r_p(\varphi)\cos x \\ r_s(\varphi)\sin x \end{pmatrix}. \quad (A1\text{-}2)$$

Since reflectance measurements are independent of the amplitude ($E_0 = 1$) (see Eqs. 2-22, 2-23) the reflectance is obtained after building the square of the absolute value of Eq. A1-2:

$$\left| \begin{pmatrix} r_p \cos x \\ r_s \sin x \end{pmatrix} \right|^2 = r_p(\varphi) r_p^*(\varphi) \cos^2 x + r_s(\varphi) r_s^*(\varphi) \sin^2 x =$$

$$= R_p(\varphi) \cos^2 x + R_s(\varphi) \sin^2 x. \quad (A1\text{-}3)$$

The statistical variation of Eq. A1-3 can be described by:

$$\iint Z(\varphi) R_p(\varphi) Z(x) \cos^2 x + Z(\varphi) R_s(\varphi) Z(x) \sin^2 x \, d\varphi dx =$$

$$= \iint Z(\varphi) R_p(\varphi) Z(x) \cos^2 x \, d\varphi dx + \iint Z(\varphi) R_s(\varphi) Z(x) \sin^2 x \, d\varphi dx =$$

$$\int_{\varphi-\alpha}^{\varphi+\alpha} Z(x) R_p(x) dx \int_{\varphi-\alpha}^{\varphi+\alpha} Z(x) \cos^2 x \, dx + \int_{\varphi-\alpha}^{\varphi+\alpha} Z(x) R_s(x) dx \int_{\varphi-\alpha}^{\varphi+\alpha} Z(x) \sin^2 x \, dx = R_p^{eff}. \quad (A1\text{-}4)$$

Since the integration steps in Eq. A1-4 are independent of each other, the effective reflectance can be expressed by ($\varphi \to x$):

$$R_p^{eff} = \int_{\varphi-\alpha}^{\varphi+\alpha} Z(x) R_p(x) dx \int_{\varphi-\alpha}^{\varphi+\alpha} Z(x) \cos^2 x \, dx + \int_{\varphi-\alpha}^{\varphi+\alpha} Z(x) R_s(x) dx \int_{\varphi-\alpha}^{\varphi+\alpha} Z(x) \sin^2 x \, dx. \quad (A1\text{-}5)$$

The first factor in each term of Eq. A1-5 results from angle divergence of the incident light beam in the plane of incidence while the second factor represents the effect of depolarization.

Although the identical distribution function Z affects both, accuracy of the angle of incidence and accuracy of the state of polarization, the effect on the measured data is of different magnitude. This is illustrated by Fig. A1-2. Here, parabola-like reflectance values for a silicon sample (photon energy 2.48 eV) are shown as derived from ellipsometry data (solid curve) [95]. This curve is compared to calculated values according to Eq. A1-5. The calculated curves were determined separately for light with imperfect polarization (dotted curve) and with divergence of the angle of incidence (dashed dotted line).

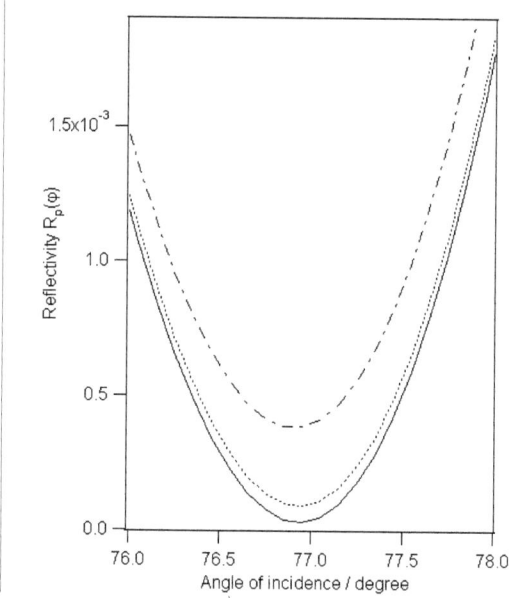

Fig. A1-2: Illustration of the effect of light beam divergence. The straight line represents the reflectance behavior around the Brewster angle of silicon at $\varphi_B \approx 76.9°$ as determined from ellipsometry data [94]. The dotted curve shows the effect of depolarized light only, the dashed-dotted curve the corresponding effect of divergence of the angle of incidence.

As a remark, optical data determined by ellipsometry are generally less affected by imperfections as discussed above. The angle of incidence in typical ellipsometric measurements is chosen some tenths of a degree off the Brewster angle [91]. Therefore, light beam divergence results in light intensities both slightly higher and lower than measured at the principal angle. Errors therefore easily cancel out while the imperfection of the polarization state contributes less to the total error as shown by Fig. A1-2.

Multi-layer analysis of BAA data is carried out, according to the preceding analysis, by comparison of theoretical R_p values, calculated from literature values or deduced from other experiments, and after transformation to *effective* BAA data, i.e. after calculation of Eq. A1-5. Since a pair of data points is determined during BAA measurements (φ_B and $R_p(\varphi_B)$), two unknown parameters can be calculated in a corresponding multi-layer analysis. Typical layer stacks comprise, however, several layers of unknown optical constants and thicknesses. Therefore, additional independent measurements are necessary in order to reduce the number of unknown quantities. This (combinatorial) approach was exemplified in section 3.1 and 3.2. A corresponding flow diagram of the applied computational procedure is finally shown in Fig. A1-3. The procedure is based on polynomial fits of both experimental BAA curves and theoretical curves. A standard self-consistent least-square routine is applied in several iterations in order to calculate missing parameters.

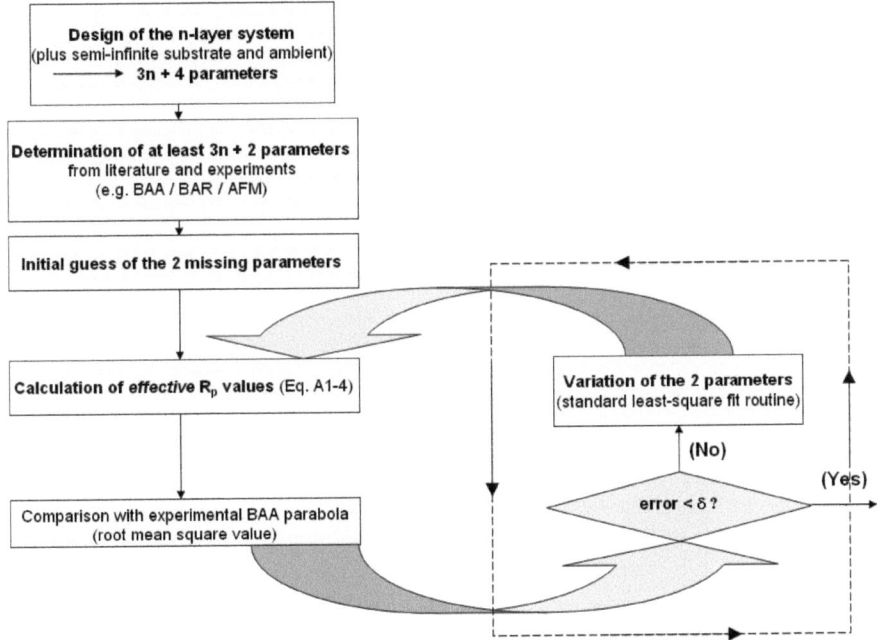

Fig. A2-1: Flow diagram of the numerical BAA multi-layer analysis. After reducing the number of unknown parameters, two parameters can be calculated. Following an initial guess of the parameters, a loop is applied in order to determine self-consistently the missing parameters by a standard least-square-fit routine. This routine evaluates experimental and theoretical parabola-like curves. A pre-defined error value, δ, terminates the loop.

A.2 Numerical procedure for simulation of the stress-induced propagation of fractal micro- and nanostructures

Fractal etch structures grow continuously with time as illustrated by Fig. 3-76a and b. The propagation velocity appears to be nearly constant for structures with high symmetry. In section 3.3.4 and 3.3.5, the oxide distribution across the surface was assumed to invoke and maintain the structure propagation. For constant propagation velocities,

$$v_x = v_y = const \quad \forall (x,y,t) \;, \tag{A2-1}$$

a homogeneous surface oxide is assumed, i.e. there exist an equilibrium of anodic oxide formation and etching for each individual branch of a fractal structure.

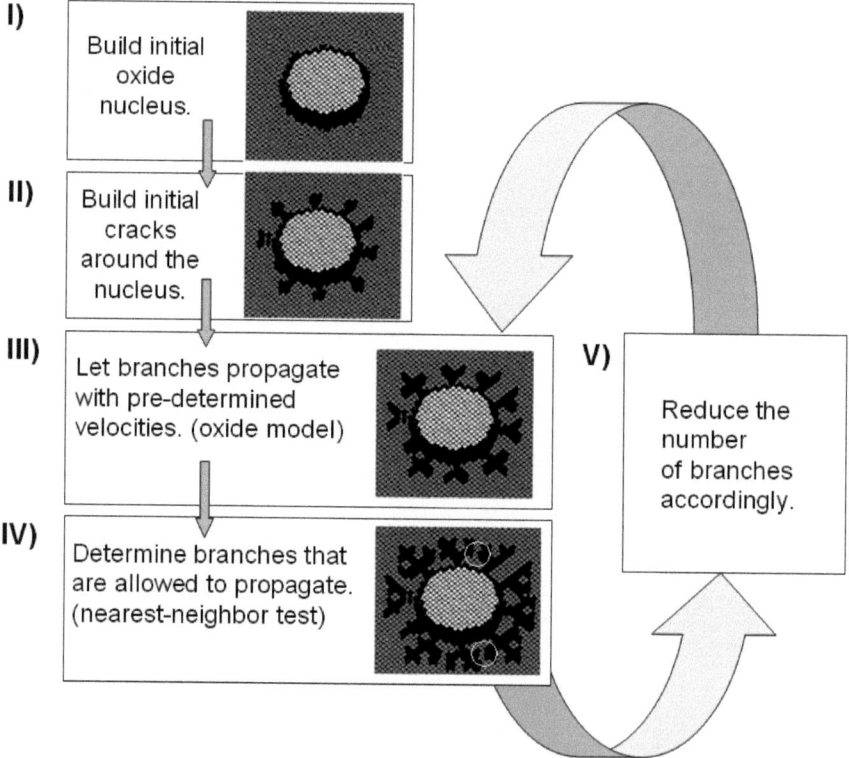

Fig. A2-1: Flow diagram for the numerical simulation procedure. The computer instructions together with the resulting surface topographies are indicated for the five principle steps. After initial crack formation, a computational loop (steps III through V) determines branches that are allowed to propagate. Branches which are too close to each other (see white circles in step IV) are excluded from further propagation.

A computer flow diagram is presented in Fig. A2-1, illustrating the steps from initial crack formation (step I), first crack propagation (step II) and repeated determination of allowed propagation directions (steps III through V).

If N denotes the initial number of cracked sites, the number of possible propagation directions in the next step is 3N provided that propagation towards the circular oxide in the center and backward-movement are not allowed. This assumption corresponds to experimental findings, detailed in section 3.2.2. The repeated addition of new cracked sites determines the set of branches, M(i), visible on the model surface after the i-th iteration of the procedure. Generally, the sets M(i) are characterized by an increasing number of elements, i.e.

$$|M(0)| < |M(1)| < |M(2)|... < |M(i-1)| < |M(i)| < ... \qquad (A2-2)$$

Step four in the scheme of Fig. A2-1, however, prevents branches being too close to each other from further growth. If there is at least one cracked site $r(x+k_1, y+k_2)$, with $(k_1)^2 + (k_2)^2 < U^2$, in the vicinity of a propagating tip with coordinates $r(x,y)$, then $r(x,y)$ is not allowed to propagate. This restriction models the stress energy release between individual parts of one or more fractal structures, illustrated by Fig. 3-38b. The number $U \in \mathbb{R}$ determines therefore the (pre-defined) minimal distance between two structures or individual branches of a single structure and is related to the strain energy stored in the near-surface region of the silicon substrate.

Since branches are eliminated by step IV, the sets M(i) differ not only in size but also with respect to their contained elements, i.e. there exists a $i \in \mathbb{N}$ such that

$$M(0) \subseteq M(1) \subseteq M(1)...M(i-1) \subseteq M(i) \not\subset M(i+1). \qquad (A2-3)$$

Eqs. A2-2 and A2-3 express that the number of branches is increasing while some of the branches fall out and others are added. Although the resulting size of the fractal structure can be estimated, the final distribution of individual branches cannot analytically be predicted.

The *graphical code representation* allows for rough assessment of the structure propagation of regular and random structures (see Fig. A2-2) and illustrates the code used in the corresponding computer program. Here, the tip of a branch is indicated as point at the center of the depicted square. Arrows indicate the possible directions of tip propagation prior to the determination of allowed directions in step IV of Fig. A2-1. For the case of high-symmetry structures on a square-grid, the graphical scheme expresses that all possible directions have the same probability 1. Therefore, no preferential direction is obtained after summation of the individual vectors.

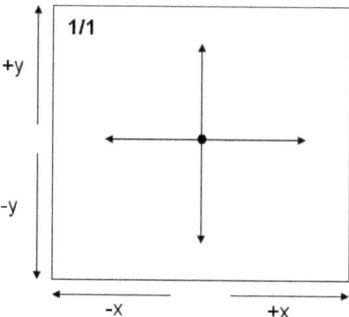

Fig. A2-2: Graphical code representation of possible directions that a branch is allowed to propagate in the presence of isotropic stress gradients. Individual vectors add up to $\vec{0}$, i.e. no preferential propagation direction is given.

Inhomogeneous oxide layers were assumed to cause random branching. Oxide-induced stress gradients change with time and branch propagation occurs with varied main directions. In this case, Eq. A2-1 has to be rewritten as:

$$v_x(x,y,t) \neq v_y(x,y,t) \neq const. \tag{A2-4}$$

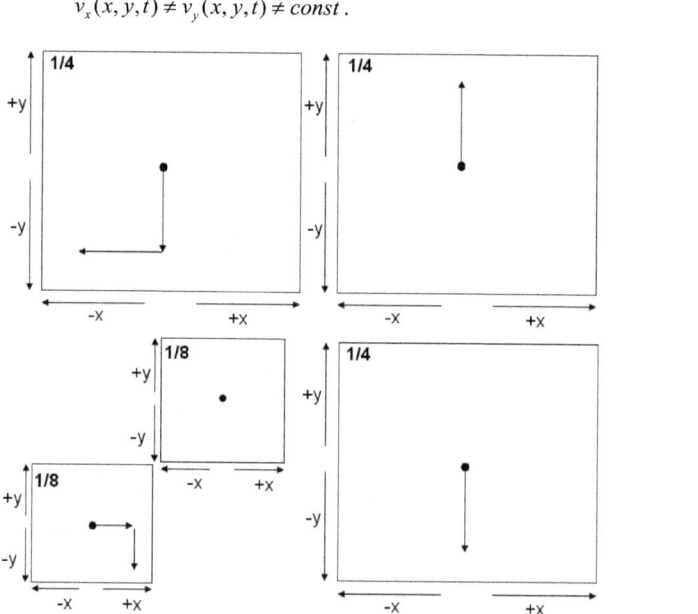

Fig. A2-3: Graphical code representation for the case of random branching in the presence of oxide layers with time-dependent local thickness. Each tip of a branch can proceed, according to the probability indicated in the upper-left corner, in different directions. Varied propagation velocities are considered by possible 0, 1 or 2 propagation steps per iteration. The respective probabilities are roughly indicated by the size of the individual squares.

Fig. A2-3 indicates the (weighted) propagation directions, determined by probability numbers in the respective upper left-hand corner of the square schemes. Variation of the propagation velocity is considered by the varied number of steps which is carried out in a single iteration cycle of the computer program. According to the scheme, no propagation step, one propagation step or two propagation steps are possible, depending on the respective probability.

The weighted addition of the individual vectors yields the most probable propagation direction although the actual distribution is not predictable. This is illustrated by Fig. A2-4, where the distribution of branches of a random structure is shown. Additionally, directions of allowed propagation (solid arrows with weighted length) and the resulting average direction (dashed arrow) are indicated as pre-defined by the scheme of Fig. A2-3.

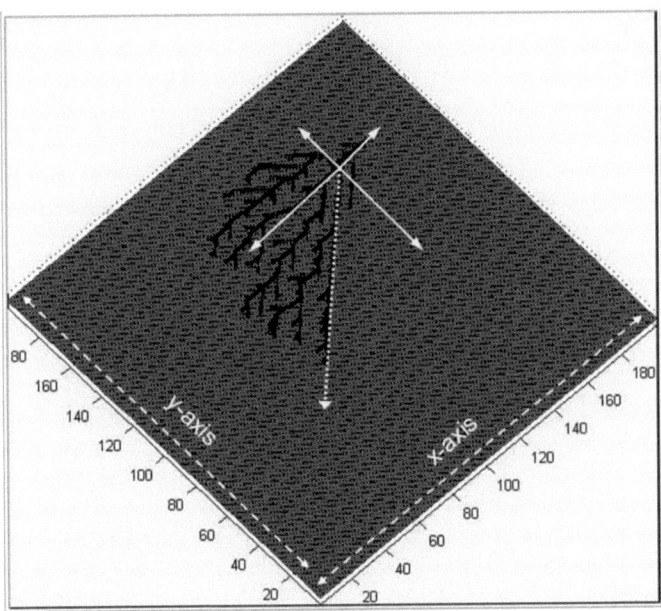

Fig. A2-4: Random structure propagation on a square grid according to the pre-defined directions and probabilities indicated in Fig. A2-3. Solid arrows illustrate allowed propagation directions. The average direction is indicated by a dashed arrow.

Fig. A2-4 illustrates that the distribution of branches is generally not predictable. Only one branch follows accurately the direction indicated by the sum vector (dashed arrow).

Acknowledgments

I am greatly indebted to Prof. Dr. H.J. Lewerenz, Prof. Dr. D. M. Kolb and Prof. Dr. J. Reif who made it possible to complete this work in a stimulating atmosphere.

I am very thankful for the collaboration with all my colleagues at the Helmholtz Zentrum Berlin who, each of them in his or her respective field, provided assistance and advice.

Particularly, the support given by Michael Kanis and Thomas Stempel-Pereira is acknowledged. Many valuable results were achieved and new ideas were developed working together with them.

Successful experiments at the SoLiAS system at Bessy II would not have been possible without the work of the collaborative research group (CRG) which I appreciate very much.

Special thanks are owed to Prof. Dr. Christian Pettenkofer and Wolfgang Bremsteller for their work at the PEEM microscope. It was especially satisfying that a theoretical approach to the understanding of fractal structure propagation could be verified by these experiments.

Financial support was granted by the Deutsche Forschungsgemeinschaft since 2005 which is greatly acknowledged. An application for prolongation was accepted in 2007 (project Nr. LE 1192/4-1/2).

References

[1] C.J. Frosch and L. Derrick, J. Electrochem. Soc., 104 (1957) 547. US Pat. 3,025,289 and 3,064,167 (1962).

[2] D. Kahng and M.M. Atalla, DRC, Pittsburgh (1960); US Patent No. 3,102, 230 (1963).

[3] A. M. Stoneham, J. L. Gavartin, A. L. Shluger, J. Phys.: Condens. Matter 17 (2005) S2027.

[4] K. Mistry, C. Allen, C. Auth, B. Beattie, D. Bergstrom, M. Bost, M. Brazier, M. Buehler, A. Cappellani, R. Chau*, C.-H. Choi, G. Ding, K. Fischer, T. Ghani, R. Grover, W. Han, D. Hanken, M. Hattendorf, J. He, J. Hicks, R. Heussner, D. Ingerly, P. Jain, R. James, L. Jong, S. Joshi, C. Kenyon, K. Kuhn, K. Lee, H. Liu, J. Maiz, B. McIntyre, P. Moon, J. Neirynck, S. Pae, C. Parker, D. Parsons, C. Prasad, L. Pipes, M. Prince, P. Ranade, T. Reynolds, J. Sandford, L. Shifren, J. Sebastian, J. Seiple, D. Simon, S. Sivakumar, P. Smith, C. Thomas, T. Troeger, P. Vandervoorn, S. Williams, K. Zawadzki, *A 45nm Logic Technology with High-k+Metal Gate Transistors, Strained Silicon, 9 Cu Interconnect Layers, 193nm Dry Patterning, and 100% Pb-free Packaging*, International Electron Devices Meeting (IEDM 2007).

[5] S. E. Lyshevski, *Nano- and Microelectromechanical Systems*, CRC Press LLC (2001):

[6] A. Korkin, J. C. Greer, G. Bersuker, Phys. Rev. B 73 (2006) 165312.

[7] N. V. Nguyen, D. Chandler-Horowitz, P. M. Amirtharaj and J. G. Pellegrino, Appl. Phys. Lett. 64 (1994) 2688.

[8] T. Emoto, K. Akimoto, Y. Ishikawa, A. Ichimiya, A. Tanikawa, Thin Solid Films 369 (2000) 281.

[9] S. M. Sze, Kwok K. Ng, *Physics of Semiconductor Devices*, John Wiley & Sons, Third Edition (2007).

[10] *Properties of Crystalline Silicon*, R. Hull (Ed.), INSPEC London (1999).

[11] C. Kittel, *Introduction to Solid State Physics*, John Wiley & Sons, Eighth Edition 200x.

[12] *Handbook of Semiconductor Silicon Technology*, W. C. O'Mara, R. B. Herring, L. P. Hunt (Eds.), Noyes Publications (1990).

[13] C. R. Helms, B. E. Deal (Eds.), *The Physics and Chemistry of SiO_2, and the Si-SiO_2 Interface 2*, Plenum Press, New York (1993).

[14] T. Aoyama, T. Yamazaki, T. Ito, J. Electrochem. Soc. 143 (196) 2280.

[15] W.-S. Liao, S.-C. Lee, J. Appl. Phys. 80 (1996) 1171.

[16] J. Stumper, *Dynamische Grenzflächenprozesse am Silizium/Elektrolyt Kontakt*, PhD Thesis, Technische Universität Berlin (1989).

[17] V. Parkhutik, Solid-State Electronics 45 (2001) 1451.

[18] F. Yahyaoui, Th. Dittrich, M. Aggour, J.-N. Chazalviel, F. Ozanam, J. Rappich, J. Electrochem. Soc. 150 (2003) B205.

[19] M. Morita, T. Ohmi, E. Hasegawa, M. Kavakami, K. Suma, Appl. Phys. Lett. 55 (1989) 562.

[20] M. Morita,T. Ohmi, E. Hasegawa, M. Kavakami, M. Ohwada, J. Appl. Phys. Lett. 68 (1990) 1272.

[21] L. Brügemann, R. Bloch, W. Press, P. Gerlach, J. Phys.: Condens. Matter 2 (1990) 8869.

[22] K.-O. Ng, D. Vanderbilt, Phys. Rev. B 59 (1999) 10132.

[23] T. Yasaka, M. Takakura, S. Miyazaki, M. Hirose, Mat. Res. Symp. Soc. 22 (1991) 225.

[24] K. Ohishi, T. Hattori, Jpn. J. Appl. Phys. 33 (1994) L675.

[25] F. J. Himpsel, F. R. McFeely, A. Taleb-Ibrahimi, J. A. Yarmoff, Phys. Rev. B 38 (1988) 6084.

[26] H. Flietner, Mat. Sci. Forum 185 (1995) 73.

[27] E. Hasegawa, A. Ishitani, K. Akimoto, M. Tsuiji, N. Ohta, J. Electrochem. Soc. 142 (1995) 273.

[28] Y. Ishikawa, M. Kosugi, M. Tabe, J. Appl. Phys. 89 (2001) 1256.

[29] L. Lai, K. J. Hebert, E. A. Irene, J. Vac. Sci. Technol. B 17 (1999) 53.

[30] Y. Yamashita, Y. Nakato, H. Kato, Y. Nishioka, H. Kobayashi, Appl. Surf. Sci. 117 (1997) 176.

[31] A. Borghesi, B. Pivac, A. Sassella, A. Stella, J. Appl. Phys. 77 (1995) 4169.

[32] Yuhai Tu, J. Tersoff, Phys. Rev. Lett. 84 (2000) 4393.

[33] L.C. Feldman, in: Y.J. Chabal (Ed.), Fundamental Aspects of Silicon Oxidation, Springer-Verlag, Berlin, 2001.

[34] A. Bongiorno, A. Pasquarello, M.S. Hybertsen, L.C. Feldman, Phys. Rev. Lett. 90 (2003) 186101-1.

[35] S. Thompson, N. Anand, M. Armstrong, C. Auth, B. Arcot, M. Alavi, P. Bai, J. Bielefeld, R. Bigwood, J. Brandenburg, M. Buehler, S. Cea, V. Chikarmane, C. Choi, R. Frankovic, T. Ghani, G. Glass, W. Han, T. Hoffmann, M. Hussein, P. Jacob, A. Jain, C. Jan, S. Joshi, C. Kenyon, J. Klaus, S. Klopcic, J. Luce, Z. Ma, B. Mcintyre, K. Mistry, A. Murthy, P. Nguyen, H. Pearson, T. Sandford, R. Schweinfurth, R. Shaheed, S. Sivakumar, M. Taylor, B. Tufts, C. Wallace, P. Wang, C. Weber, M. Bohr, IEDM (2002) 61.

[36] S. E. Thompson, S. Suthram, Y. Sun, G. Sun, S. Parthasarathy, M. Chu, T. Nishida,

Future of Strained Si/Semiconductors in Nanoscale MOSFETs, IEDM Tech. Digest (2006) 1.

[37] O. Marty, T. Nychyporuk, J. de la Torre, V. Lysenko, G. Bremond, D. Barbier, Appl. Phys. Lett. 88 (2006) 101909.

[38] V. Lysenko, D. Ostapenko, J.-M. Bluet, P. Regregny, M. Mermoux, A. Boucherif, O. Marty, G. Grenet, V. Skryshevsky, G. Guillot, Phys. Status Solidi A (2009) DOI 10.1002/pssa.200881103 (to be published).

[39] Martin H. Sadd, *Elasticity – Theory, Applications and Numerics*, Elsevier (2005).

[40] L. B. Freund, *Dynamic Fracture Mechanics*, Cambridge University Press (1998).

[41] T. Emoto, K. Akimoto, A. Ichimiya, Surf. Sci. 438 (1999) 107.

[42] L. D. Landau, E. M. Lifshitz, *Theory of Elasticity*, Pergamon Press, second (revised) Edition (1970).

[43] A. Bongiorno, A. Pasquarello, Phys. Rev. B 24 (2000) R16326.

[44] F. Giustino, A. Bongiorno, A. Pasquarello, J. Phys.: Condens. Matter 17 (2005) S2065.

[45] R. A. Marcus, J. Chem. Phys. 24 (1956) 966 and 979.

[46] R. A. Marcus, Annu. Rev. Chem. 15 (1964) 155.

[47] H. Gerischer, Z. Phys. Chem. N. F. 26 (1960) 223.

[48] H. Gerischer, Z. Phys. Chem. N. F. 27 (1961) 48.

[49] H. Gerischer in *The CRC Handbook of Solid State Electrochemistry*, P. J. Gellings, H. J. M. Bouwmeester (Eds.), CRC Press (1997).

[50] V. S. Bagotsky, *Fundamentals of Electrochemistry*, John Wiley & Sons (2006).

[51] J. Koryta, J. Dvořák, L. Kavan, *Principles of Electrochmistry*, second edition, John Wiley & Sons (1993).

[52] R. Memming, *Semiconductor Electrochemistry*, Wiley-VCH, Weinheim (2001).

[53] C. H. Hamann, W. Vielstich, *Elektrochemie*, Wiley-VCH, Weinheim (2005).

[54] H. Gerischer, J. Electroanal. Chem. Interfacial Electrochem. 58 (1975) 263.

[55] F. Lohmann, Z. Naturforsch. 22A (1967), 843.

[56] H. Angermann, *Chemische Konditionierung der Silicium-Oberfläche: Präparation und Charakterisierung von Wasserstoff-terminierten und naßchemisch oxidierten Si(111)- und Si(100)-Oberflächen*, PhD Thesis, Freie Universität Berlin (1999).

[57] P. Allongue, V. Kieling, H. Gerischer, Electrochimica Acta 40 (1995) 1353.

[58] M. A. Hines, Y. J. Chabal, T. D. Harris, A. L. Harris, J. Chem. Phys. 101 (1994) 8055.

[59] N. Tomita, S. Adachi, J. Electrochemical Society 149 (2002) G245.

[60] S. Wiggins, *Introduction to Applied Nonlinear Dynamical Systems and Chaos*, Springer

Verlag (1990).

[61] D. K. Arrowsmith, C. M. Place, *Dynamical Systems – Differential equations, maps and chaotic behaviour*, Chapman & Hall (1992).

[62] S. H. Strogatz, *Nonlinear Dynamics and Chaos: with applications to physics, biology, chemistry and engineering*, Addison-Wesley Publishing Company (1994).

[63] J. Guckenheimer, P. Holmes, *Nonlinear Oscillations, Dynamical Systems and Bifurcations of Vector Fields*, (1983).

[64] K. Krischer, *Nonlinear Dynamics in Electrochemical Systems* in *Advances in Electrochemical Science and Engineering*, R. C. Alkire, D. M. Kolb (Eds.), Wiley-VCH (2002).

[65] D. R. Turner, J. Electrochem. Soc., 105 (1958) 402.

[66] V. Lehmann, J. Electrochem. Soc. 143 (1996) 1313.

[67] H. J. Lewerenz, J. Electroanal. Chem. 351 (1993) 159.

[68] J. Grzanna, H. Jungblut, H. J. Lewerenz, J. Electronal. Chem. 486 (2000) 181.

[69] J.-N. Chazalviel, F. Ozanam, J. Electrochem. Soc. 139 (1992) 2501.

[70] J. Carstensen, R. Prange. G. S. Popkirov, H. Föll, Appl. Phys. A 67 (1998) 459.

[71] J. Grzanna, H. Jungblut, H. J. Lewerenz, J. Electronal. Chem. 486 (2000) 190.

[72] A. Uhlir, Bell. Syst. Tech. J., 35 (1956) 333.

[73] L. T. Canham, Appl. Phys. Lett. 57 (1990) 1046.

[74] C. Pickering, M. I. J. Beale, D. J. Robbins, P. J. Pearson, R. Greef, J. Phys. C: Solid State Physics 17 (1984) 6535.

[75] C. Pickering, M. I. J. Beale, D. J. Robbins, P. J. Pearson, R. Greef, Thin Solid Films 125 (1985) 157.

[76] A. G. Cullis, L. T. Canham, Nature 353 (1991) 335.

[77] K.W. Kolasinski, Current opinion in Solid State and Materials Science 9 (2005) 73–83.

[78] T.F. Young, I.W. Huang, Y.L. Yang, W.C. Kuo, I.M. Jiang, T.C. Chang, C.Y. Chang, Appl. Surf. Sci. 102 (1996) 404.

[79] N. Happo, M. Fujiwara, M. Iwamatsu, K. Horii, Jpn. J. Appl. Phys. 37 (1998) 3951.

[80] M. Wesolowski, Phys. Rev. B 66 (2002) 205207.

[81] V. M. Aroutiounian, M. Zh.Ghoolinian, and H. Tributsch, *Appl. Surf. Sci.*, 162 (2000) 122.

[82] L. N. Aleksandrov, P. L. Novikov, Thin Solid Films 330 (1998) 102.

[83] V. Lehman, J. Electrochem. Soc. 140 (1993) 2836.

[84] S. Rönnebeck, J. Carstensen, S. Ottow, H. Föll, Electrochem. Solid-State Lett. 2 (1999)

126.

[85] M. Christophersen, *Untersuchungen zur Makroporenbildung in Silizium und deren technologischen Nutzung*, PhD Thesis, Christian-Albrechts-Universität zu Kiel (2002).

[86] J. C. Claussen, J. Carstensen, M. Christophersen, S. Langa, H. Föll, Chaos 13 (2003) 217.

[87] J. Carstensen, M. Christophersen, H. Föll, Mater. Sci. Eng. B69 (2000) 23.

[88] K. Kopitzki, *Einführung in die Festkörperphysik*, B. G. Teubner Stuttgart (1993).

[89] W. Nolting, *Grundkurs Theoretische Physik 3 – Elektrodynamik*, Springer-Verlag (2002).

[90] A. Sommerfeld, H. Bethe, *Elektronentheorie der Metalle*, Heidelberger Taschenbuch, vol. 19, Springer-Verlag (1967).

[91] H. Fujiwara, *Spectroscopic Ellipsometry – Principles and Applications*, John Wiley & Sons (2007).

[92] H. J. Lewerenz, N. Dietz, Appl. Phys. Lett., 59 (1991) 1470.

[93] N. Dietz, H. J. Lewerenz, Appl. Phys. Lett. 60 (1992) 2403.

[94] H. J. Lewerenz, N. Dietz, J. Appl. Phys., 73 (1993) 4975.

[95] T. Yasuda and D. E. Aspnes, Appl. Opt. 33 (1994) 7435.

[96] N. Dietz, *Charakterisierung von Halbleitern für photovoltaische Anwendungen mit Hilfe der Brewster-Winkel-Spektroskopie*, PhD Thesis, Technische Universität Berlin (1991).

[97] R. M. A. Azzam, N. M. Bashara, *Ellipsometry and Polarized Light*, North Holland, Amsterdam (1989).

[98] D.J. Bergman, Phys. Rep. C 43 (1978) 377.

[99] D.J. Bergman and D. Stroud, in: Solid State Physics, Vol. 46, Eds. H. Ehrenreich and D. Turnbull (Academic Press, San Diego, 1992) p. 148.

[100] W. Theiß in: *Festkörperprobleme, Advances in Solid State Physics*, Vol. 33, R. Helbig (Ed.), Vieweg, Braunschweig (1994).

[101] J.C. Maxwell Garnett, Philos. Trans. R. Soc. London 203 (1904) 385.

[102] D.A.G. Bruggeman, Ann. Phys. 24 (1935) 636.

[103] S. Hüfner, *Very High Resolution Photoelectron Spectroscopy*, Springer-Verlag (2007).

[104] M. Henzler, W. Göpel, *Oberflächenphysik des Festkörpers*, B. G. Teubner Stuttgart (1994).

[105] W. Nolting, *Grundkurs Theoretische Physik 5/2: Quantenmechanik - Methoden und Anwendungen*, Springer-Verlag Berlin Heidelberg New York (2002).

[106] W. E. Spicer, Phys Rev Vol 112 (1958) 114.

[107] J. F. Watts, J. Wolstenholme, *An Introduction to Surface Analysis by XPS and AES*, John Wiley & Sons (2003).

[108] J. D. Jackson, *Classical Electrodynamics*, John Wiley & Sons (1962).

[109] C. Gerthsen, H. O. Kneser, H. Vogel, *Physik*, Springer-Verlag (1992).

[110] A. Hofmann, *The Physics of Synchrotron Radiation*, Cambridge University Press (2004).

[111] *Surface and Thin Film Analysis: Principles, Instrumentation, Applications*, H. Bubert, H. Jennet (Eds.), Wiley-VCH Verlag (2002).

[112] M. P. Seah, W. A. Dench, Surf. Interface Anal. 1 (1979) 2.

[113] E. Brüche, Z. Phys. 86 (1933) 448.

[114] O.H. Griffith, G.F. Rempfer, Adv. Opt. Electron. Microsc. 10 (1987) 269.

[115] H. H. Rotermund, S. Jakubith, S. Kubala, A. von Oertzen, G. Ertl, J. Elec. Spec. Rel. Phen. 52 (1990) 811.

[116] C. M. Schneider and G. Schönhense, Rep. Prog. Phys. 65 (2002) R1785.

[117] J. M. Garguilo, *Electronic Transition Imaging of Carbon Based Materials: The Photothreshold of Melanin and Thermionic Field Emission from Diamond*. PhD Thesis, North Carolina State University (2007).

[118] G. F. Rempfer, OH Griffith, Ultramicroscopy 27 (1989) 273.

[119] M. Kotsugi, et al., Rev. Sci. Instr. 74 (2003) 2754.

[120] G. Binnig, H. Rohrer, Helvetica Physica Acta 55 (1982) 726.

[121] G. Binnig, H. Rohrer, Surf. Sci. 126 (1983) 236.

[122] G. Binnig, C. F. Quate, C. Gerber, Phys. Rev. Lett. 56 (1986) 930.

[123] A. Foster, W. Hofer, *Scanning Probe Microscopy*, Springer Science + Business Media, LLC (2006).

[124] *Scanning Probe Microscopy Training Notebook*, Version 3.0, Digital Instruments, Veeco Metrology Group (2000).

[125] F. London. Trans. Faraday Soc. 33 (1937) 8.

[126] P. J. W. Debye. Phys. Z., 21 (1920) 178.

[127] P. J. W. Debye. Phys. Z., 22 (1921) 302.

[128] Communications Physical Laboratory, Leyden, Holland (1912).

[129] H. C. Hamaker. Physica, 4 (1937) 1058.

[130] D. Sarid, *Scanning Force Microscopy*, Oxford University Press, New York (1991).

[131] B. V. Derjaguin, V. M. Muller, Y. P. Toporov, J. Colloid Interf. Sci. 53 (1975) 314.

[132] *Applied Scanning Probe Methods II*, B. Bhushan, H. Fuchs, Springer-Verlag Berlin

Heidelberg (2006).

[133] P. J. Goodhew, J. Humphreys, R. Beanland, *Electron Microscopy and Analysis*, Taylor & Francis, New York (2001).

[134] M. Knoll, Z. Tech. Phys. 16 (1935) 467.

[135] M. von Ardenne, Zeitschrift für Physik 108 (1938) 553.

[136] M. von Ardenne, Zeitschrift für technische Physik 19 (1938) 407.

[137] G. S. Higashi, Y. J. Chabal, G. W. Trucks, K. Raghavachari, Appl. Phys. Lett. 56 (1990) 656.

[138] G. S. Higashi, R. S. Becker, Y. J. Chabal, A. J. Becker, Appl. Phys. Lett. 58 (1991) 1656.

[139] P. Allongue, C. H. de Villeneuve, S. Morin, R. Boukherroub, D. D.M. Wayner, Electrochimica Acta 45 (2000) 4591.

[140] P. Dumas, Y. J. Chabal, J. Vac. Sci. Technol. A 10 (1992) 2160.

[141] M. Copel, R. J. Culbertson, R. M. Tromp, Appl. Phys. Lett. 65 (1994) 2344.

[142] M. Niwano, J.-I. Kageyama, K. Kurita, K. Kinashi, I. Takahashi, N. Miyamoto, J. Appl. Phys. 76 (1994) 2157.

[143] S. Ye, T. Saito, S. Nihonyanagi, K. Uosaki, P. B. Miranda, D. Kim, Y.-R. Shen, Surf. Sci. 476 (2001) 121.

[144] S.-E. Bae, M.-K. Oh, N.-K. Min, S.-H. Paek, S.-I. Hong, C.-W. J. Lee, Bull. Korean Chem. Soc. 25 (2004) 1822.

[145] J. Fu, H. Zhou, J. Kramar, R. Silver, S. Gonda, Appl. Phys. Lett. 82 (2003) 3014.

[146] G. J. Pietsch, U. Köhler, M. Henzler, J. Appl. Phys. 73 (1993) 4797.

[147] M. Nakamura, M.-B. Song, M. Ito, Electrochimica Acta 41 (1996) 681.

[148] M. Niwano, Y. Kondo, Y. Kimura, J. Electrochem. Soc. 147 (2000) 1555.

[149] J. Flidr, Y.-C. Huang, T. A. Newton, M. A. Hines, J. Chem. Phys. 108 (1998) 5542.

[150] M. A. Hines, Annu. Rev. Phys. Chem. 54 (2003) 29.

[151] M. A. Gosálvez, K. Sato, A. S. Foster, R. M. Nieminen, H. Tanaka, J. Micromech. Microeng. 17 (2007) S1.

[152] P. Dumas, Y. J. Chabal, P. Jakob, Surf. Sci. 269 (1992) 867.

[153] *Fundamental Aspects of Silicon Oxidation*, Y. J. Chabal (Ed.), Springer-Verlag Berlin Heidelberg (2001).

[154] D. Hojo, N. Tokuda, K. Yamabe, Thin Solid Films 515 (2007) 7892.

[155] F. J. Himpsel, J. L. McChesney, J. N. Crain, A. Kirakosian, V. Pérez-Dieste, N. L. Abbott, Y.-Y. Luk, P. F. Nealey, D. Y. Petrovykh, J. Phys. Chem. B 108 (2004) 14484.

[156] T. Sekiguchi, S. Yoshida, K. M. Itoh, Phys. Rev. Lett. 95 (2005) 106101.

[157] V. B. Svetovoy, J. W. Berenschot, M. C. Elwenspoek, J. Micromech. Microeng. 17 (2007) 2344.

[158] H. Sakaue, Y. Taniguchi, Y. Okamura, S. Shingubara, T. Takahagi, Appl. Surf. Sci. 234 (2004) 439.

[159] P. Jakob, P. Dumas, Y. J. Chabal, Appl. Phys. Lett. 59 (1991) 2968.

[160] G. S. Hsiao, J. A. Virtanen, R. M. Penner, Appl. Phys. Lett. 63 (1993) 1119.

[161] R. Zhu, E. Pan, P. W. Chung, X. Cai, K. M. Liew, A. Buldum, Semicond. Sci. Technol. 21 (2006) 906.

[162] A. Roy Chowdhuri, Dong-Un Jin, J. Rosado, C. G. Takoudis, Phys. Rev. B 67 (2003) 245305.

[163] J. H. Ouyang, X. S. Zhao, T. Li, and D. C. Zhang, J. Appl. Phys. 93 (2003) 4315.

[164] K. J. Hebert, S. Zafar, E. A. Irene, R. Kuehn, T. E. McCarthy, E. K. Demirlioglu, Appl. Phys. Lett. 68 (1995) 266.

[165] Y. Wang, E. A. Irene, J. Vac. Sci. Technol. B18 (2000) 279.

[166] S. Iwata, A. Ishizaka, J. Appl. Phys. 79 (1996) 6653.

[167] S. R. Kasi, M. Liehr, S. Cohen, Appl. Phys. Lett. 58 (1991) 2975.

[168] M. Niwano, K. Kurita, Y. Takeda, N. Miyamoto, Appl. Phys. Lett. 62 (1993) 1003.

[169] H. J. Lewerenz, T. Bitzer, J. Electrochem. Soc.139 (1992) L21.

[170] Q.-Y. Tong, Q. Gan, G. Fountain, G. Hudson, P. Enquist, Appl. Phys. Lett. 87 (2004) 2762.

[171] H.J. Lewerenz, M. Aggour, C. Murrell, M. Kanis, H. Jungblut, J. Jakubowicz, P. A. Cox, S. A. Campbell, P. Hoffmann, D. Schmeisser, J. Electrochem. Soc. 150 (2003) E185.

[172] M. Grundner, R. Schulz, Conf. Proc. No. 167, American Vacuum Soc. 4 (1988) 329.

[173] Jolly, F. Rochet, G. Dufour, C. Grupp, A. Taleb-Ibrahimi, J. Non-Cryst. Solids 280 (2001) 150.

[174] Carniato, J.-J. Gallet, F. Rochet, G. Dufour, F. Bournel, S. Rangan, A. Verdini, L. Floreano, Phys. Rev. B 76 (2007) 085321.

[175] V. Lehmann, U. Gösele, Adv. Materials 4 (1992) 114.

[176] *Porous Silicon*, Z.C. Feng, R. Tsu (Eds.), World Scientific, Singapore (1994).

[177] A. G. Cullis, L. T. Canham, P. D. J. Calcott, J. Appl. Phys. 82 (1997) 909.

[178] *Properties of porous silicon*, L. Canham (Ed.), INSPEC, London (1997).

[179] C. W. Corbett, D. J. Shereshevsky, I. V. Verner, Phys. Status Solidi A 147 (1995) 81.

[180] V. P. Parkhutik, J. M. Albella, J. M. Martinez-Duart, J. M. Gomez-Rodriguez, A. M. Baro, V. I. Shershulsky, Appl. Phys. Lett. 62 (1993) 366.
[181] V. Dubin, Surf. Sci. 274 (1992)82.
[182] P. Allongue, C. Henry de Villeneuve, L. Pinsard, M. C. Bernard, Appl. Phys. Lett. 67 (1995) 691.
[183] M. I. J. Beale, J. D. Benjamin, J. Uren, N. G. Chew, A. G. Cullis, J. Cryst. Growth 73 (1985) 622.
[184] H. Gerischer, P. Allongue, V. Kieling, Ber. Bunsenges. Phys. Chem. 97 (1993) 753.
[185] V. Lehmann, U. Gösele, Appl. Phys. Lett. 58 (1991) 856.
[186] I. M. Young, M. I. Beale, J. D. Benjamin, Appl. Phys. Lett. 46 (1985) 1133-5.
[187] D. Bellet, G. Dolino, Thin Solid Films 276 (1995) 1.
[188] H. Sugiyama, O. Nittono, J. Cryst. Growth 103 (1990) 156.
[189] V. Lysenko, D. Ostapenko, J.-M. Bluet, P. Regregny, M. Mermoux, A. Boucherif, O. Marty, G. Grenet, V. Skryshevsky, G. Guillot, Phys. Status Solidi A, DOI 10.1002/pssa.200881103.
[190] M. L. Lee, E. A. Fitzgerald, M. T. Bulsara, M. T. Currie, A. Lochetefeld, J. Appl. Phys. 97 (2005) 011101.
[191] A. Nakajima, T. Futatsugi, K. Kosemura, T. Fukano, N. Yokoyama, Appl. Phys. Lett. 71 (1997) 353.
[192] S. Tiwari, F. Rhana, H. Hanafi, A. Harstein, E. F. Crabbé, K. Chen, Appl. Phys. Lett. 68 (1996) 1377.
[193] T. Nychyporuk, V. Lysenko, B. Gautier, D. Barbier, Appl. Phys. Lett. 86 (2005) 213107.
[194] T. Nychyporuk, V. Lysenko, B. Gautier, D. Barbier, J. Appl. Phys. 100 (2006) 104307.
[195] Th. Dittrich, S. Rauscher, V. Yu. Timoshenko, J. Rappich, I. Sieber, H. Flietner, H. J. Lewerenz, Appl. Phys. Lett. 67 (1995) 1134.
[196] J. Jakubowicz, H. Jungblut, H.J. Lewerenz, Electrochimica Acta 49 (2003) 137.
[197] K. Skorupska, J.Jakubowicz, H. Jungblut, H.J. Lewerenz, Superlattices and Microstructures 36 (2004) 211.
[198] H. Jungblut, J. Jakubowicz, H.J. Lewerenz, Surf. Sci. 597 (2005) 93.
[199] H.J. Lewerenz, M. Aggour, C. Murrell, J. Jakubowicz, M. Kanis, S.A. Campbell, P.A. Cox, P. Hoffmann, H. Jungblut, D. Schmeißer, J. Electroanal. Chem. 540 (2003) 3.
[200] K. W. Kosalinski, J. Electrochem. Soc. 152 (2005) J99; 153 (2006) L28.
[201] H.J. Lewerenz, H. Jungblut, S. Rauscher, Electrochimica Acta, 45 (2000) 4615.

[202] V. Lehmann, J. Electrochem. Soc. 140 (1993).

[203] V. Lehmann, *Electrochemistry of Silicon: Instrumentation, Science, Materials and Applications*, Wiley-VCH Verlag GmbH (2002).

[204] H.J. Lewerenz, J. Jakubowicz, H. Jungblut, C. R. Chimie 9 (2006) 289.

[205] J. Rappich, V. Y. Timoshenko, R. Würz, T. Dittrich, Electrochimica Acta 45 (2000) 4629.

[206] R. Outemzabet, M. Cherkaoui, N. Gabouze, F. Ozanam, N. Kesri, J.-N. Chazalviel, J. Electrochem. Soc. 153 (2006) C108.

[207] P. D. Bourke, Computers and Graphics, 30 (2006) 134.

[208] B. B. Mandelbrot, *The Fractal Geometry of Nature*, Freeman, New York (1982).

[209] K. Falconer, *Fractal Geometry – Mathematical Foundations and Applications*, John Wiley & Sons (2003).

[210] J. A. Kaandorp, *Fractal Modelling – Growth and Form in Biology*, Springer-Verlag (1994).

[211] Edgar E. Peters, *Fractal Market Analysis*, John Wiley & Sons (1994).

[212] D. Sornette, *Critical Phenomena in Natural Sciences – Chaos, Fractals, Selforganization and Disorder: Concepts and Tools*, Springer-Verlag (2000).

[213] L. Spanos, E. A. Irene, J. Vac. Sci. Technol. A 12 (1994) 2646.

[214] L. Spanos, Q. Liu, E. A. Irene, T. Zettler, B. Hornung, J. J. Wortman, J. Vac. Sci. Technol. A 12 (1994) 2653.

[215] V. M. Aroutiounian, M. Zh.Ghoolinian, H. Tributsch, *Appl. Surf. Sci.*, 162 (2000) 122.

[216] M. Wesolowski, Phys. Rev. B 66 (2002) 205207.

[217] T. Qiu, X. L. wu, Y.F. Mei, P.K. Chu, G.G. Siu, Appl. Phys. A 81 (2005) 669.

[218] H. Föll, M. Christophersen, J. Carstensen, G. Hasse, Mat. Sci. Eng. R 39 (2002) 93.

[219] A. G. Muños and M. M. Lohrengel, J. Solid State Electrochem. 6 (2002) 513.

[220] L. Saraf, D. R. Baer, Z. Wang, J. Young, M. H. Engelhard, S. Thevuthasan, Surf. Interface Anal. 37 (2005) 555.

[221] K. Jacobi, M. Gruyters, P. Geng, T. Bitzer, M. Aggour, S. Rauscher, H. J. Lewerenz, Phys. Rev. B 51 (1995) 5437.

[222] T. Homma, Thin Solid Films 278 (1996) 28.

[223] M. P. López-Sancho, F. Guinea, E. Louis, J. Phys. A: Math. Gen. 21 (1988) L1079.

[224] P. Meakin, G. Li, L.M. Sander, E. Louis, and F. Guinea, J. Phys. A: Math. Gen. 22 (1989) 1393.

[225] O. Pla, F. Guinea, E. Louis, G. Li, L. M. Sander, H. Yan, P. Meakin, Phys. Rev. A. 42

(1990) 3670.

[226] V. Parkhutik, Mater. Sci..Engineer. B 58 (1999) 95.

[227] V. Parkhutik, Solid-State Electronics 43 (1999) 1121.

[228] S. Langa, J. Carstensen, M. Christophersen, K. Steen, S. Frey, I. M. Tiginyanu, H. Föll, J. Electrochem. Soc. 152 (2005) C525.

[229] V. Lehmann, H. Föll, J. Electrochem. Soc. 137 (1990) 653.

[230] M. Lublow, *Brewster-Winkel-Spektroskopie an Sauerstoff- implantiertem Gallium Phosphid und vergleichende spektroskopische Methoden*, Master Thesis, Brandenburgische Technische Universität Cottbus (2003).

Die VDM Verlagsservicegesellschaft sucht für wissenschaftliche Verlage abgeschlossene und herausragende

Dissertationen, Habilitationen, Diplomarbeiten, Master Theses, Magisterarbeiten usw.

für die kostenlose Publikation als Fachbuch.

Sie verfügen über eine Arbeit, die hohen inhaltlichen und formalen Ansprüchen genügt, und haben Interesse an einer honorarvergüteten Publikation?

Dann senden Sie bitte erste Informationen über sich und Ihre Arbeit per Email an *info@vdm-vsg.de*.

Sie erhalten kurzfristig unser Feedback!

VDM Verlagsservicegesellschaft mbH
Dudweiler Landstr. 99   Telefon +49 681 3720 174
D - 66123 Saarbrücken   Fax   +49 681 3720 1749

www.vdm-vsg.de

Die VDM Verlagsservicegesellschaft mbH vertritt

Printed by Books on Demand GmbH, Norderstedt / Germany